Life Cycles

An evolutionary approach
to the physiology of
reproduction, development and ageing

Life Cycles :

An evolutionary approach to the physiology of reproduction, development and ageing

PETER CALOW
Department of Zoology,
University of Glasgow

LONDON

CHAPMAN AND HALL

A Halsted Press Book
John Wiley & Sons, Inc., New York

First published 1978
by Chapman and Hall Ltd
11 New Fetter Lane, London EC4P 4EE

© *1978 P. Calow*

Typeset by C. Josée Utteridge-Faivre and printed in Great Britain by
Richard Clay (The Chancer Press) Ltd, Bungay, Suffolk

ISBN 0 412 21510 1 (cased edition)
ISBN 0 412 21520 9 (Science Paperback)

Distributed in the U.S.A. by Halsted Press,
a Division of John Wiley & Sons, Inc., New York

Library of Congress Cataloging in Publication Data

Calow, Peter.
  Life cycles.

  Bibliography: p.
  1. Life (Biology)   2. Reproduction.   3. Developmental biology.   4. Aging.   I. Title.
QH501.C36   577   78-15401
ISBN 0-470-26474-8

Contents

Preface   *page* viii

PART ONE INTRODUCTION   1

1   **Life's logic**   2
1.1 Nature is cyclical   2
1.2 There are two ways of replicating a system   4
1.3 Evolution of a self-replicating system is more exciting   5
1.4 Life as a communication channel   6
1.5 Some simple population genetics   7
1.6 Molecular details of biological systems   11
1.7 Cellular details of biological systems   13
1.8 The organism   16
1.9 Summary   17

PART TWO GROWING   19

2   **Life as an energy transforming system**   20
2.1 Life as an open system   20
2.2 The partitioning of energy in organisms   21
2.3 Problems in trying to assess efficiency   24
2.4 Some estimates of conversion efficiency in heterotrophs   25
2.5 Photosynthetic efficiency   29
2.6 Metabolic adaptation   30
2.7 Metabolic adaptability   33

3   **Life as a cellular system**   37
3.1 The cell cycle and growth   39
3.2 Differentiation and growth   41
3.3 Nutrient supply and cell division   45

4 **Life as a dynamic steady-state** *page* 47
4.1 Tissues in flux 47
4.2 Molecules in flux 48
4.3 Cells in flux 50
4.4 Control of cell turnover 53

5 **On the adaptive significance of growing** 56
5.1 There are two ways to get bigger 56
5.2 Is there any benefit in adiposity and obesity? 60

PART THREE REPRODUCTION 65

6 **How organisms reproduce** 66
6.1 Mitosis and meiosis 66
6.2 Sexual dimorphism 69
6.3 Ferilization in animals 71
6.4 Events leading to fertilization in plants 73
6.5 Why sex? 75
6.6 Why two sexes? 83
6.7 Conclusions 86

7 **Quantitative aspects of reproduction** 87
7.1 When to reproduce and by how much? 87
7.2 Whether to reproduce all at once or repeatedly? 90
7.3 The cost of reproduction 91
7.4 Examples of reproductive strategies 96
7.5 Reproductive adaptability 100
7.6 The size and number of gametes 104

PART FOUR AGEING 107

8 **The ageing process** 108
8.1 What happens to organisms that do not meet an unnatural death? 108
8.2 The manifestation of ageing in the individual 110
8.3 Single or multiple theories of ageing? 112
8.4 Ageing by accident or design? 113
8.5 A hypothesis based on the random, non-programmed accumulation of molecular damage 115
8.6 Conclusions 123

Contents   vii

9   The cycle reversed   *page*   124
9.1   Degrowth in triclads   124
9.2   Physiological evidence for rejuvenation in triclads 127
9.3   Actuarial evidence for rejuvenation   129
9.4   Rejuvenation by asexual fission   133
9.5   Rejuvenation by sexual reproduction   133

PART FIVE SUMMARY AND CONCLUSIONS   135

References   137

Index of organisms   152

Subject index   156

Preface

As time progresses, biology becomes more and more fragmented and specialized and it becomes increasingly difficult to see how all the disparate facts fit together. It is completely proper that biologists should have sought to reduce complex biological wholes into their parts, and it is natural that studies on the products of this reduction should have diverged from more holistic studies on evolution and ecology. Yet the biological parts, what they do and how they are organized are products of an evolutionary process which fits organisms for life in particular ecological circumstances. Physiology, developmental biology, ecology and evolutionary biology must not be allowed to grow too far apart, therefore, because all these disciplines and the way their subject matters interact are crucial to understanding organisms — and it is this, it seems to me, which is the fundamental goal of the biological sciences.

This book has been written in the spirit of unification and synthesis. It is, in a sense, a general biology of the organism — not, however, of organisms as static unchanging systems, but of organisms as dynamic entities which progress through a definite cycle of events from birth to maturity. The central theme, therefore, will be the life cycle, and the book is organized around the three main phases which are characteristic of all life cycles; growth (Part II), reproduction (Part III) and ageing (Part IV). Part I is an introductory section on the general cyclical nature of both the animate and inanimate worlds and Part V is a general summary which attempts to define the central problem for organismic biology.

I have written the book for anyone who feels a need for a more wide-ranging, interdisciplinary approach to the study of organisms, whether they be undergraduates on biology courses, medical students in their introductory years or people already in biological research. The book does assume some prior knowledge of biology but not, I hope, a great deal.

I cannot pretend that it covers all it ought, or that it treats what is covered in a balanced fashion. The bias reflects my own interests and all I can hope is that the reader finds those subjects that are treated more thoroughly than others as absorbing as I obviously do.

In presenting the arguments I have inevitably referred to a wide variety of organisms, unicellular and multicellular, plant and animal. In an effort to put these in their taxonomic places without overburdening the text with too much taxonomic detail, I have included a short 'index of organisms' at the end of the book, which groups those organisms referred to in the text into broad taxonomic categories. This may be used as a stepping-stone into more specialist taxonomic works.

Glasgow 1978 P.C.

Errata

The Publishers apologize for the presence in this book of the errors listed below, for which they accept full responsibility.

Page 1   a line has been omitted between lines 2 and 3. It should read: 'I ask two kinds of question: 'How does it take place?' and 'Why does it'
Line 9 should be deleted.

Page 5   line 2 from foot, for 'though is is' read 'though it is'.

Page 8   line 19, for 'though and equilibrium' read 'though an equilibrium'.
line 24, for 'to some extent to the effects' read 'to some extent the effects'.

Page 16   last line, for 'lead to evolution' read 'led to the evolution'.

Page 17   line 5, for 'Profiera, Oelenterata' read 'Porifera, Coelenterata'.

Page 27   line 6 from foot, for 'Remongton' read 'Remington'.

Page 29   line 15, for 'homeotaxis' read 'homeostasis'.
line 20, for 'metazoa and less' read 'metazoa are less'.
line 21, for 'absorphan' read 'absorption'.

Page 42   line 14, for 'present' read 'presence'.

Page 45   line 5, 'Section 3.1' should be in parentheses.

Page 49   line 20, for 'Schinke' read 'Schimke'.

Page 61   line 3, for 'could' read 'should'

Page 62   line 6, for 'of the equivalents' read 'of the energy equivalents'.

Page 63   line 7, for 'substistence' read 'subsistence'.

Page 66   line 14, for 'termatodes' read 'trematodes'.

Page 70   line 7, for 'theses' read 'these'.

Page 76   line 21, for 'argues' read 'argued'.

Page 82   line 2, for 'occurence' read 'occurrence'.

Page 84   line 21, for $[R + (m -f)] + [R + (m -f)] = 2R + 2(m -f)$
read $[R -(m + f)] + [R -(m + f)] = 2R -2(m + f)$
line 24, for 'to spend more energy' read 'more energy to spend'.
lines 32–33, for 'hermaphoditism' read 'hermaphroditism'.

Page 86   heading 6.7 should read 'Conclusions'.

Page 93   line 25, for 'process' read 'processes'.

Page 104   line 9, for 'thus' read 'they'
line 10, for 'that' read 'but'.
line 21, for 'thing s' read 'things',

Page 108   line 3, for 'at the same' read 'at the same time'.

Page 113   line 8 from foot, for 'means' read 'mean'.

Page 116   line 2 from foot, for 'Wulf Cutler' read 'Wulf and Cutler'.

Page 119   line 18, for 'W/w' read 'W/V'.

Page 120   line 11, for 'Ycas' read 'Yčas'.

Page 127   Equation 9.3 should read: $A_f = A_\infty (1 - e^{-(k/3)(t - t_0)})^2$

Page 133   line 18, for 'along' read 'why'.

Page 134   line 2, for 'procedes' read 'proceeds'.

Page 158   for '*Drosophila* melangogaster' read '*Drosophila melanogaster*'.

Life Cycles by P. Calow, published by Chapman and Hall.

Part one Introduction

This book is about the cycle of events which enables the products of reproduction themselves to reproduce. About each part of this life cycle take place in the way it does?' The 'how-questions' seek answers in terms of proximate, physiological and biochemical causes and usually form the subject matter of textbooks in developmental biology. On the other hand, the 'why-questions' seek answers in terms of the more ultimate processes of selection and adaptation and form the subject matter of most textbooks in evolutionary biology. My aim, however, has not been most textbooks in Evolutionary Biology. My aim, however, has not been to present new developmental facts nor to elaborate new evolutionary theories but to bring both subjects to bear on key problems of organismic biology. The rationale behind this method of approach to the study of organisms is that ontogeny can only be fully appreciated as a product of phylogeny and that phylogeny can only be fully understood in terms of the way it operates on ontogenies, not just on specific parts of the life cycle like the adult. In an analysis of this kind, another essential ingredient is ecology, for it is tacitly understood that organisms are adapted in terms of their life cycles to where they live. Hence, three subjects are brought together in the pages that follow – the biology of development, evolution and ecology. I begin, though, in a very general way by considering the adaptive significance of the cyclical nature of the biological processes themselves and by attempting to reconcile the conflict between this cycle, life's cycle, and the unidirectionality predicted by the second law of thermodynamics. In so doing my aim is to crystallize the major features of the logic of organismic and evolutionary biology and thus to establish a framework for the critical, in-depth study of the main components of the life cycle. This analysis follows in Parts II to IV where I consider in turn, growth, reproduction and ageing.

1

1 Life's logic

1.1 Nature is cyclical

At its most basic, the second law of thermodynamics states that all
reactions proceed towards increased entropy. But if this was always the
case there could be no cycles in nature, just a continuous, unidirectional
trend from order to disorder; and for this reason the second law has
sometimes been called the barb in time's arrow (Blum, 1955). Ordered
and organized things do appear to be common in the world around us,
however, and on Earth living beings have always raised something of a
problem for the supposed universality of the entropic trend. At least
part of the solution to this puzzle lies in the fact that the second law
applies only to closed systems through which there is no overall flow of
energy or matter. Here, particles are supposed to interact in a random
fashion such that the general organization of the system tends to some
base state which is the one of least order. The universe as a whole may
or may not be considered as a closed system, depending on one's own
cosmological philosophy, but in any event, parts of it may clearly be
treated as open to the through-flow of energy. What happens within
these streams of energy is that particles of matter are activated, direction
is given to the otherwise random processes, and in this way patterns of
energy and matter are generated that would otherwise have been highly
improbable. Furthermore, energy flowing out of the open system is
invariably more disordered than the energy flowing in, so that though
the entropy trend within the energy-corridor may be reversed, that for
the whole system, inside and out, is in keeping with the predictions of
the second law. Even within the open system most of the complexes fall
back to the norm within a short time so that cycles of generation and
degeneration, rather than unidirectionality, become a predominant
feature of the real, natural world and a continual inflow of energy is

required to preserve the order and organization. Basically therefore, the processes occuring throughout nature are cyclical, involving a continuous flux between order and disorder, generation and degeneration. Nevertheless, some configurations appear to persist within this flux for considerable lengths of time and this occurs in one of two ways. Either particles are brought together in extremely stable states by chance, or the same pattern is formed over an over again. Persistence in the first place is due to the property of stability, whereas persistence in the second place is due to the process of repeatability. Stability leads to a static persistence whereas repeatability leads to a dynamic persistence.

A good example of what is meant by stability is the step-by-step formation of carbon from helium. The encounter of two hydrogen nuclei gives rise to helium because the 'helium-type organization' is more stable than any other configuration of hydrogen nuclei. If helium nuclei then collide in pairs the resulting association is unstable and breaks apart in less than 10^{-12} of a second. However, if in this split-second of time a third helium nucleus collides with the pair, the three hold together in an extremely stable configuration representing the nucleus of carbon. All carbon atoms now in existence have been formed in this way.

Dynamic persistence, the repeated formation of the same system, can itself occur in one of two ways. On the one hand there can be fortuitous replacement when the same configuration reappears at different times but when each appearance is an independent happening. For this to occur, each system must have a high probability of formation and cannot, in consequence, be very complex. As systems, clouds are common and yet sooner or later they are liable to break apart. The reason why clouds appear to persist is that they are continuously being formed afresh. Alternatively, other systems persist in a dynamic way by replication. Provided a system can copy itself before it is broken down, a complex organization can be kept in existence. In this case, therefore, there is a definite connection between the appearance of one system and the re-appearance of replicas. Furthermore, since replication can result in the formation of numerous copies even a system which forms only once (i.e. has a low probability of formation) can become very common. This sort of behaviour occurs in crystal formation and also in the reproduction of living organisms.

1.2 There are two ways of replicating a system

Organisms die, but before they do they leave replicas which persist in
their place. For long-term persistence the process of self-replication must
be sufficiently stable to ensure that the progeny retain just those features
that enable them to last long enough to re-replicate, and they must also
be sufficiently flexible to ensure that the organization of the system
remains in tune with an inevitably changing environment. Now there are
two ways of copying an already existing system, but only one of them is
consistent with these requirements.

Consider an example. John and Jim are friends. John has a model
aeroplane which Jim would like to build for himself. If Jim was clever
enough he might ask John for the model and then proceed to copy it
direct. If Jim copied the model blindly there would be three possible
sources of error: a) he would copy the mistakes that John had originally
made in building the model; b) he would probably make some mistakes
himself; c) he would copy damage, accumulated accidentally in the course
of the life of the plane; for example, a broken wing, a bent propellor
and so on. If somebody then copied Jim's version, somebody copied the
copy and so on, the errors would inevitably build up such that each
successive model looked and acted less like the original.

There is an alternative strategy. Jim could ask for a copy of the
original plans and build his model from that. Here, there would be two
potential sources of error: a) in copying out the plans and b) in building
the model. However, the most significant source of error, accident
through involvement in some kind of physical work, would be eliminated
because unlike the actual plane, the plans are never required to do very
much; they are read, not used in any active sense and though they may
become 'dog-eared' the chances of them being seriously damaged are
very much less than those of the plane. Hence, as the plans were handed
from friend to friend the form of the product would change only
slightly so the replicated planes would retain their plane-like character
for a larger number of 'generations' than if they had been copied
directly.

From the experience of Jim and John it would seem that a system can
only persist by direct replication if it is stable and inert. Otherwise
replication must involve the transmission of some kind of plans. In
keeping with these views crystals, which are both stable and inert, can
replicate by a direct copying process (Cairns-Smith, 1971) whereas
organisms, which are extremely active, meta-stable systems, replicate
by the transmission of plans embodied within their hereditary material.

The secret behind dynamic systems which persist by replication, therefore, is a bipartite make-up; an inert, stable set of instructions which specify a product capable, at least in the self-replicating case, of being able in turn to re-copy and re-transmit the plans. It is exactly in this way, of course, that living things are organized, consisting as they do of a genotype, the plans, and a phenotype, the product.

1.3 Evolution of a self-replicating system is more exciting

It is often assumed that evolution by natural selection can take place only in biological systems, but this is not completely true. Cosmic reactions continuously generate a large variety of chemical systems and of these, unstable ones are *selected* against as matter and energy are funnelled into other, stable systems. Only stable chemicals persist. The chemical world as a whole therefore evolves into a stable state by selection. Thus stable chemical systems, like carbon and iron, *evolve* out of unstable precursors. Persistence here, of course, is in terms of one lasting organization of matter and if that system perishes, the world must wait until another stable configuration develops in its place. There is no repeatability, no passing-on of experience and in consequence, the tempo of change is slow and never very 'adventurous'.

Transmission through replication inevitably leads to an increased tempo in evolution and makes possible an altogether more 'adenturous' process. Provided the design specifications are sufficiently protected there is continuity, since a system which happens to exist long enough to re-replicate can persist even if the parent then meets with a mortal accident. Hence, replication acts like a ratchet which prevents ordered systems slipping back into disorder and it also acts like a photocopier spreading, by multiplication, a particular design feature. Of course, slight alterations to the design specifications caused by copying errors will bring about alterations in the product and the character of the replicating system will change. Coupled with this, it is obvious that in a finite environment only those systems which can obtain resources (matter and energy) will re-replicate. In other words only those systems which are best fitted to replicate themselves under particular conditions of resource availability will persist. Because there will be a continuous tendency for mistakes to turn up in the copying process there will be a continuous generation of variety in replicating systems and because of competition there will be a tendency for better and better adapted systems to evolve. The process may look as though is is directed by an intelligence but in fact the designing force is natural selection.

1.4 Life as a communication channel

Systems which persist by replication pass on only a small amount of
energy and matter from parent to offspring. Replication occurs instead
through the transmission of information; through, that is, the transmission
of a programmed set of instructions, the plans, which specify a replica.
Therefore, lineages produced by replication can be considered as communication channels in which individual organisms receive a message from
the preceding generation and, after amplification, pass it on to the next
generation. The message passed on in this way consists of a simple command
'transmit me', which is associated with subsidiary instructions specifying
how re-transmission should be effected under particular ecological
conditions.

In one way or another the re-transmission of genetic instructions
depends on activity; chemical, mechanical and even mental. However, the
experience with the model aeroplane taught us that the plans must not
be too active if they are to be protected against accidental damage. Hence
the inactive genotype must express itself through an active agent, the
phenotype, which moves, 'nourishes' and protects the message prior to
re-replication. A second system of communication, from genotype to
phenotype, is therefore associated with the first and is concerned with
catalysing the basic instruction 'to replicate'.

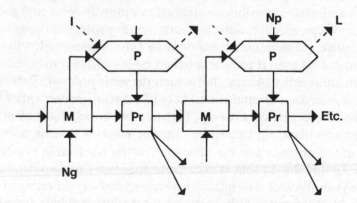

Fig. 1.1 Cybernetic model of an asexual system. M = genome; P = phenotype;
Pr = 'genetic printer'; I = energy input; Np = noise influencing phenotype;
Ng = noise influencing genotype. Information flow is represented by solid
lines and energy flow by broken lines. (With permission from Calow 1977a,
Adv. Ecol. Res. **10**. Copyright by Academic Press Inc. (London) Ltd.)

A very crude and general model of a genetic system is illustrated in Fig. 1.1.
Here M represents the genotype and P the phenotype; solid lines represent

information and broken lines energy flow. Note that information flows from *M* to *P* but not in the other direction. This is because even a successful phenotype will bear the mark of its activity in the form of accidentally acquired damage. Since much of this damage, if allowed to accumulate, would work against the basic instruction 'to replicate', selection could not have allowed the reverse transmission of information from phenotype to genotype. This precludes Lamarckian evolution and is often referred to as the 'central dogma of molecular biology'. Informational and energetic disturbances on *P* (somatic mutations) may influence replication but are never transmitted. Only informational changes in *M* due to mutation, and in sexual forms, recombination, crossing over, etc. are transmissible.

1.5 Some simple population genetics

The phenotype does its job by interacting with its environment. How successful or fit it is depends on whether it survives long enough to reproduce and on how many viable offspring it leaves. Fitness is dependent not only on differential survival, therefore, but upon differential fecundity. Both of these characters, survival and fecundity, are properties of individual organisms, yet population genetics is concerned largely with specific genes and what part they play in the make-up of the total collection of genes distributed through all the individuals in the population — the gene pool as it is called. The reason for this orientation is two-fold. Firstly, it simplifies the mathematical treatment of genetical problems and secondly, it reflects in a more realistic way how the genetical system works. As already noted, the genetic model described in the last section is only an approximation as far as sexual systems are concerned, for crossing-over, independent assortment and random mating mean that whole genetic programmes are not transmitted intact but that the genes are separable and can be treated as independent units. More will be said about this in Section 6.5. With some reservations (see Kempthorne and Pollak, 1970), therefore, it should be possible to define selection and fitness unambiguously in terms of gene frequencies, and to define evolution in terms of a change in gene frequency. The aim of population genetics has been to make these ideas explicit and mathematically rigorous.

The theory of population genetics usually begins with an imaginary diploid population with the following hypothetical properties: random mating; numbers sufficiently large for statistical reliability; negligible

changes in gene frequency due to mutation and migration. Consider two
alleles X and Y (alternative genes specifying for the same character)
within this population, with relative frequencies x and y respectively
($x + y = 1$). There are, therefore, three possible genotypes, XX, XY, YY.
Let the frequency of these genotypes be $a + 2b + c$, so x is always equal
to $a + b$ and y to $b + c$. Now with random mating each individual may
mate with any other of the three genotypes with a probability that
depends on their genotype frequencies, viz.:

	XX	XY	YY males
XX	a^2	$2ab$	ac
XY	$2ab$	$4b^2$	$2bc$
YY	ac	$2bc$	c^2
females			

From this, the frequencies of XX, XY, and YY in the F_1 generation are
respectively $(a + b)^2$, $2(a + b)(b + c)$ and $(b + c)^2$. But we already know
that $a + b = x$ and $b + c = y$, so the frequencies reduce to $x^2 + 2xy + y^2$
or $(x + y)^2$. If we were to repeat the operation for any other subsequent
generation the frequency composition would remain constant throughout at $(x + y)^2$ and this important result is known as the Hardy-Weinberg
theorem. It is, in a sense, a re-statement of Mendel's first law for populations. Clearly the Hardy-Weinberg theorem is an idealization which,
like the ideal gas laws, applies only to fictitious systems and not to
anything which actually exists. For example, the theorem does not
strictly apply to sex-linked genes and here, though and equilibrium state
is achieved, it is reached only slowly. Also selection, gene imports and
exports due to migration and gene alterations due to recurrent mutation
will alter the outcome of the simple predictions. Nevertheless, by using
the Hardy-Weinberg theorem as a fixed starting point it is possible to
quantify to some extent to the effects of mutation, migration and selection.
For such a wide-ranging treatment see any textbook on population
genetics (e.g. Mettler and Gregg, 1969; Cook, 1971); in this section we
concentrate only on the influence of selection on the 'Hardy-Weinberg
mechanics'.

Let us return to the population with X and Y alleles which is diploid,
very large and in which there is no migration and little mutation.
Suppose the zygotes of X contribute x_0 gametes to the zygotes of the
next generation and the zygotes of Y contribute y_0 gametes to the
zygotes of the next generation, then the selective value (W) of X relative
to Y is given by x_0/y_0. (If it were possible to determine the density of X

or Y in recurrent generations it would be possible to calculate an absolute selective value but this is often difficult in practice). Now in the next generation:

$$x_1 = \frac{W x_0}{y_0 + W x_0}$$

and the change in frequency is:

$$\Delta x = \frac{W x}{y + W x} - x$$

$$= \frac{-xy\,(1-W)}{y + W x}$$

The equilibrium position ($\Delta x = 0$) depends largely on the numerator for if it is zero the value of the denominator is of little account. The numerator is zero if x or y are zero (i.e. if only one allele is present) or if $W = 1$ (i.e. there is no selection). In this case, therefore, selection leads to the fixation of whichever allele is the most favoured. Clearly, this simple conclusion is greatly complicated by the partitioning of genes between homozygotes and heterozygotes (e.g. X between XX and XY) because the association of genes in the heterozygote may, dependent on the nature and extent of dominance, shelter one gene from the influence of selection or subject the other to selection of an increased intensity. In considering this more complex situation it is useful to substitute $1-C$ for W; C being the coefficient of selection. When $x < y$, C is positive and X is selected against relative to Y; when $x > y$, C is negative and X is selectively favoured.

Consider first the simple case where XX has a coefficient of selection of C relative to YY and the heterozygote XY has a coefficient of $C/2$ relative to YY. Then it can be shown (Alexander, 1967) that the number of X to Y genes will reduce e (2.7) times every $2/e$ generations. Of course, this is only a simplified approximation since the coefficients of selection often have the following relationships: $C_{XX} = C_{XY} \neq C_{YY}$ (i.e. X is dominant to Y) or sometimes $C_{XX} \neq C_{XY} \neq C_{YY}$ (i.e. X and Y interact to produce an effect different to either of their separate effects), and the more general equation describing the rate of change of x is:

$$\Delta x = \frac{xy\,(C_{YY}\,y - C_{XX}x) - C_{XY}\,(1 - 2x)}{1 - C_{YY}y^2 - 2C_{XY}xy - C_{XX}x^2}$$

This equation predicts a sigmoid change in gene frequency with time. If X is rare it increases in frequency rapidly at first but then slows down as the equilibrium state is reached. If X is dominant the initial change is fast but the eventual fixation slow whereas if X is recessive the reverse is true. The equilibrium position ($\Delta x = 0$) depends crucially on the relative values of C_{XX}, C_{XY} and C_{YY}. When $C_{XX} < C_{XY} > C_{YY}$ (heterozygote advantage) or $C_{XX} > C_{XY} < C_{YY}$ (heterozygote disadvantage) an equilibrium may be reached at which both alleles are retained in the population. Otherwise there is either no equilibrium position at all or one involving just one allele (the one with the negative C). Returning to the simple assumption of no dominance, however, it is possible to use C to make approximate calculations on the kinds of rates of replacement or spread of characters that are likely in a population. Clarke and Sheppard (1966) found that the non-melanic form of the peppered moth, *Biston betularia,* had a coefficient of selection of + 0.6 relative to the melanic form in industrial areas. Applying the simple assumptions outlined above it can be calculated that over two generations the ratio of non-melanic to melanic moths in the population would reduce by about threefold. Hence it is easy to see how melanism quickly spread through the population after industrialization. I shall return to this simple treatment again in Chapter 2.

The adaptive value (fitness) of any trait depends on how well it is transmitted. This, in turn, depends on mating success, fertility, survival to maturity and ultimately on the fecundity of the organism bearing the trait. There is also one other important possibility. A gene can promote the increasing frequency of replicas of itself in a population not only by programming the 'selfish' reproductive activity of the organism that bears it but also by programming the bearer to 'help' other organisms that carry the same gene to survive and reproduce themselves. This is often referred to as altruism and should clearly operate only between closely related individuals. For this reason altruism is said to evolve by kin selection and its effects are well exemplified by parental care and by the 'selfless devotion' of workers to queens in social Hymenoptera. It cannot be overemphasized, however, that kin selection will only work between individuals which have a high probability of bearing the same gene and that it does not give general licence for evolutionary explanations based on the benefit of the species or group as a whole. Indeed, if such group-selectionist views ever became acceptable it would mean that most of our current ideas about natural selection were wrong. For a more rigorous account of kin selection the reader is referred to Hamilton (1972) and for a more readable account to Dawkins (1976).

1.6 Molecular details of biological systems

Over the past twenty years it has become increasingly possible to characterize the genotype and the phenotype chemically and to explain the communication of genetic information solely in terms of physical chemistry
(Watson, 1970). We now know, for example, that the genetic instructions
are carried in coded form by nucleic acids, usually DNA. This information
is transmitted from generation to generation by replication and is transmitted within each generation via RNA, to proteins. The proteins domin-
ate the structure and function of the phenotype by being involved
directly, as building blocks, in crucial structural units and by being
involved as catalysts in all the major metabolic pathways. The system is
strongly circular in that proteins are themselves involved in catalysing
their own formation in the transcription of DNA to RNA and in the
translation of RNA into proteins. In addition, proteins are needed to
catalyse replication (Fig. 1.2).

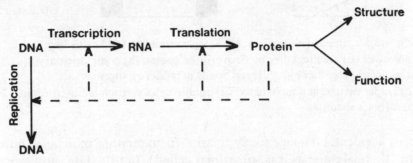

Fig. 1.2 Molecular version of Fig. 1.1. See text for further explanation.

In biochemical terms, the genotype–phenotype dichotomy can be
re-defined as the nucleic acid–protein dichotomy. Nucleic acids regulate
protein synthesis and proteins regulate metabolism. A series of RNAs
link the two systems. The genetic information on the nucleic acids is
specified by the variation of base molecules in the string of nucleotides
(Fig. 1.3a). Four bases are involved and in DNA these are: adenine,
thymine, guanine and cytosine. Complimentary strings of nucleotides
are held together by hydrogen bonds (Fig. 1.3b), the resulting ladder
arrangement is twisted like an electric flex into a double helix (Fig. 1.3c)
and this is held together by more hydrogen bonds, Van der Waals forces
and polar and ionic interaction. In eukaryotes proteins are intertwined
through and around the DNA and thus give it further support. (Fig. 1.3d).
The resulting DNA is very stable and thus protects the genetic instructions

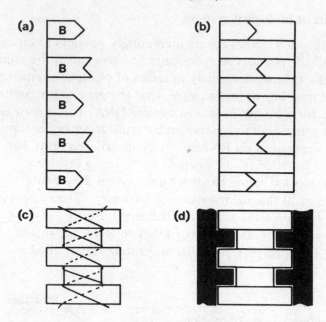

Fig. 1.3 Structure of chromatin: (a) a single strand of DNA with base molecules (B) held together by deoxyribose sugars; (b) complimentary strands held together by hydrogen bonds in ladder arrangement; (c) ladder twisted in double helix; (d) double helix surrounded by protein coat (black shading).

against accidental damage likely to arise from random molecular bombardment (thermal noise as it is sometimes called). In this state, however, with the information on the inside of the molecule, the DNA cannot do anything, so for the purpose of both replication and transcription the molecular coils must be unwound and the hidden nucleotides must be laid bare. It is at this stage that thermal noise has its most potent effect in generating somatic and genetic mutation. In some cases the resulting molecular damage can be repaired (section 8.5), in others the products may be eliminated as inviable gametes or impotent molecules and in a few cases the damage may be retained and actually improve the performance of the system.

At the phenotypic end of the genetic communication system are the proteins. These are made up of amino acids, of which twenty different sorts are involved. The properties of the proteins are dependent on the types and arrangements of these subunits within the molecule. However, the way a protein works does not depend directly on the linear sequence of molecules within it (the primary and secondary structure) but rather on the three-dimensional shape of the whole complex. By 'holding' substrate molecules in the interstices of its coils a protein can cause them to react in a particular way and at a particular rate and it is in this way that proteins exert their catalytic effect. The physical properties of the structural proteins like hair, wool, nail and so on also depend on their shape. Thermal noise has the effect of deforming the three-dimensional configuration of proteins and may thus render them impotent, and this is another way the phenotype can be disturbed. Deformed or denatured molecules are usually degraded and excreted but occasionally they may continue to do their job in a sub-standard way. In so doing, damaged proteins may produce more imperfect molecules and may amplify their own effect. Ultimately the damage expresses itself by impairing the metabolic well-being of the organism as a whole (Chapter 8).

Since four bases specify for twenty amino acids a single base molecule could not code for a single amino acid. If each of two bases coded for one amino acid then a total of 4^2 or 16 amino acids could be specified — again insufficient. However, with three bases per amino acid, 4^3 or 64 different possibilities could be specified, which is more than enough to account for the 20 amino acids. In fact it is now well-known that the genetic code is based on this triplet organization with the excess $(64-20 = 44)$ being taken up by redundancy (e.g. CCU, CCC, CCA and CCG all coding for proline) and punctuation (e.g. TAA = full stop). Of course, the transmission of information from DNA to protein is not direct since there is a complex RNA system in between. As with DNAs, though, the RNAs consist of four bases and of these only one is different; uracil replaces thymine. Hence the DNA to RNA transmission of information is straightforward and complications only arise in converting RNA information into proteins. This is mediated through a series of small, allosteric RNA molecules, the tRNAs, which are shaped such that they can couple to a specific amino acid at one end and a specific base-triplet at the other. In this way the tRNAs plug into an mRNA template and thereby bring amino acids into a specific linear association in which they can be coupled enzymatically to form specific proteins.

1.7 Cellular details of biological systems

To be effective, metabolic reactions and their controls have to be held together in physical proximity. The origin of organic systems capable of self-replication, therefore, depended on the formation of membranes

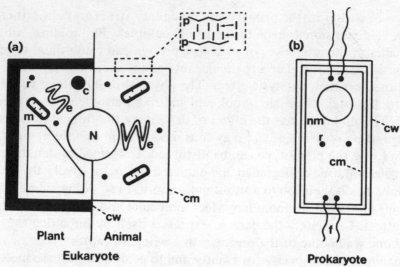

Fig. 1.4 Generalized cellular structures. c = chloroplast; cw = cell wall;
e = endoplasmic reticulum; f = flagellae; l = lipid; m = mitochondria;
n = nucleus; nm = nuclear material; p = proteins; r = ribosome.

which could surround a collection of molecules whose quality and
quantity were optimal for metabolic purposes. These metabolic packets
are the cells and most of them are around $10-20$ μm in diameter, but
there is some considerable variation. The largest known cell, the ostrich
egg, is over 100 mm in diameter whereas the smallest known cell, the
Pneumococcus bacterium, is less than 0.1 μm in diameter.

All eukaryotic cells typically consist of the genetic instructions, as
strand-like chromosomes, locked in the membrane-bound nucleus. The
metabolic apparatus is located in the surrounding cytoplasm and there is
an outer, semipermeable membrane of protein and lipid molecules
(Fig. 1.4a). Bacteria and blue-green algae, known collectively as prokaryotes,
lack a nuclear membrane and mitochondria and have a peculiar chromosomal arrangement, but otherwise conform to the basic pattern (Fig. 1.4b).
In what follows I shall be concerned almost exclusively with eukaryotes.

The DNA is wound in rather a complex, not completely understood
way into the chromosomal strands. Active sites unwind and cause the
chromosome to 'puff out' in the region of activity (Fig. 1.5). Genetic
instructions pass from the chromosomes over the nuclear membrane as
mRNA and the proteins are then fashioned on cytoplasmic particles rich
in RNA known as the ribosomes.

Metabolic reactions proceed in the cytoplasm under the direction of

Fig. 1.5 Puffs in the giant chromosomes of *Drosophila* are thought to be composed of loops like this. The hairs are the strands of mRNA being synthesized on the DNA.

the proteins. However, the energy-producing reactions take place within other sub-cellular organelles, the mitochondria. These hold the respiratory enzymes in a definite physical orientation such that the exergonic reactions can proceed in an orderly and controlled fashion. The mitochondria also contain DNA of their own as well as ribosomes, tRNA and all the other apparatus needed for protein synthesis. It is thought that this allows mitochondria to replicate themselves and to act as semi-autonomous units. Because of this apparent autonomy, one theory asserts that mitochondria are 'organisms within organisms' which entered into a symbiotic relationship with the cell at an early stage in its evolution.

As dictated by the 'central dogma' the flow of information within the cells is from nucleus to cytoplasm. However, conditions in the cytoplasm, themselves dependent on conditions in the external environment, may influence the expression of the genetic information and thus the quality and quantity of enzyme secretion. For example, *Escherichia coli,* when grown in a medium containing lactose, quickly produces an enzyme, β-galactosidase, to degrade it. This well-documented phenomenon seems to occur because lactose interacts with specialized allosteric molecules within the bacterium which in turn have a switching action on the genetic instructions. In the absence of lactose these 'switching molecules' bind to the genome and inactivate just that portion concerned with the formation of β-galactosidase. By combining with other sites on it, lactose inactivates the 'switching molecule' and thereby 'switches on' the formation of the necessary enzyme. Mechanisms of this sort keep the cell in tune with the environment in which it finds itself and prevent it from wasting time and resources on the manufacture of enzyme systems which have no immediate value.

1.8 The organism

It is not easy to come up with a satisfactory definition of the organism. Organisms are made of cells but they may also consist of just one cell as in the Bacteria and Protista. Organisms are self-replicating systems, but so are the cells within organisms. However, organisms effect replication by the conversion of resources from the outside world into themselves. No cell from a metazoan will do this unless it is propagated in culture. Hence, organisms are the units which compete for the limited resources in the environment and as such are the active units of selection.

There have been three important trends in the evolution of organismic systems: a progressive increase in size (sometimes referred to as Cope's Law) and cellularization; a progressive increase in cellular differentiation; a progressive increase in the degree of co-ordination and communication between the components of the organism.

With the appearance of each new phylum, organismic size has been pushed upwards (Chapter 5). Since the cell surface-to-volume ratio falls with increases in cell size these increases in organismic size could not have occurred by an increase in the volume of constituent cells. Instead the larger the organisms have become the more cells they have come to contain (Table 1). It is possible, in fact, that selection pressure for increased size together with physiological problems associated with a larger cell volume-to-surface ratio lead to the evolution of multicellularity in the first place. In metazoans, total cell surface grows more or less in proportion to body weight, not to body surface, so by cellularization the size of individuals could be increased without impairing the physiological processes dependent on a high surface-volume relationship (Zeuthen, 1947).

Increases in organismic size have also been correlated with increases in cellular diversification; different cells in the same organism becoming specialized to do different things. This is illustrated in Table 1. It is easy to appreciate the evolutionary rationale behind this trend. Systems concentrating on a single function are likely to be more efficient at that function than a 'Jack of all trades' because the needs of different functions often conflict. Larger systems require more resources and more efficient use of resources so that cellular diversification might be a pre-requisite for increases in organismic size.

Specialization, of course, brings with it a greater need for intercellular communication and co-operation. This has been met by the evolution of chemical messengers. Furthermore, in higher animals the need for communication has lead to evolution of complex hormonal and nervous

Table 1 Approximate number of cells and number of types of cells in animals from different 'levels of organization'.

	Approx. No. of cells	Approx. No. of cell types
Volvox	10^2	2
Placozoa	10^3	5–10
Porifera	$> 10^3$	> 10
Hydra	10^8	10–20
Triclads	10^9	100
Annelids	10^{12}	100
Insects	10^{12}	100
Humans	10^{15}	1000

systems and of neurohumoral complexes like the pituitary. Specialization has also been associated with reduced flexibility and at least one manifestation of this is that as organisms become more specialized their powers of regeneration become proportionately more restricted. In the Animal Kingdom the Profiera, Oelenterata and Turbellaria can regenerate whole bodies from small portions of tissue; the Arthropoda and Amphibia can only regenerate limbs and certain other appendages and in the higher vertebrates the regenerative capacity becomes restricted almost entirely to wound-healing.

1.9 Summary

Nature's cycle oscillates between generation and degeneration. Life's cycle involves generation, degeneration and generation again following replication. Replication occurs through the transmission of a coded design usually in individual cells or, as happens in vegetative reproduction, in a small fragment of tissue. At the same time there has been an undisputed trend in evolution towards multicellularity and large size, and with increases in size have come increases in differentiation. Hence, after transmission, germ cells and propagules have to grow and differentiate into a new, reproductively active, organism. Finally there is the obvious fact that failing death by accident, predation and disease most organismic systems degenerate and age. Consequently the life cycle of the organism consists of three major events: 'growth', reproduction and senescence (Fig. 1.6). In what follows I consider *how* each of these processes is brought about and *why*, if at all, it is programmed to occur when and how it does. The prime requirement of a successful biological system is to transmit and spread the genetic information it carries — but what

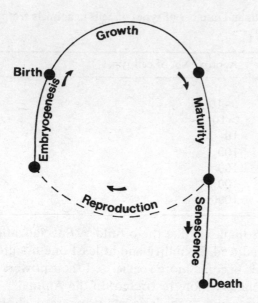

Fig. 1.6 The major events in the life cycle of an organism.

strategy it uses depends largely on the ecological circumstances in which it occurs. Looking at the relationship between ecology and organismic organization as it has been described in this chapter will be a central theme in what follows.

Part two Growing

Growth occurs when the input of nutrients to the organism is more than enough to make good the losses incurred in the dissipation processes. This excess fills out the organism through the building up of an elaborate internal organization composed of cellular subunits. Materials and energy made available for growth are used either to produce new cells, to fill out old cells or to create extracellular material like fat. The metabolic basis of growth is considered in Chapter 2 and the cellular basis in Chapter 3. Ultimately, however, these ways of expressing growth must merge since cells are made out of matter and energy and since 'biomass' is a product of cellular activity. This will be considered in the latter part of Chapter 3.

In crude terms, the generative part of the life cycle involves the patterned but not perfect reproduction of an original parental system. On top of this, there is a continuous replacement of what has already been produced. Even at steady-state size, for example, there is a turnover of molecules and cells within the organism and once again the orderliness of this replacement is ensured by a persisting genetic programme. This 'dynamic steady-state' will be examined in Chapter 4.

Finally, in Chapter 5 we turn to the evolutionary question, 'Why grow?' It turns out that the answer to this question is not at all obvious since in basic evolutionary terms the fittest organism ought to be the smallest. In fact it is only when the life cycle is judged against the requirements for survival posed by specific environmental conditions that the sense in growing very big becomes apparent.

2 Life as an energy transforming system

2.1 Life as an open system

It is indisputable that living systems at all levels are thermodynamically open. Globally, life originated and continues to persist in a stream of energy flowing from a source, the sun, to a sink, the rest of the universe. There is an input of sunlight and an output of heat energy (Fig. 2.1).

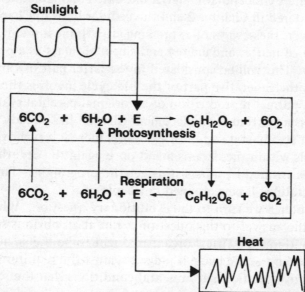

Fig. 2.1 The circulation of matter within and the flow of energy through the biosphere.

The sunlight drives an endergonic reaction which builds organic molecules from inorganic precursors. This light-powered synthesis, photosynthesis, depends on the absorption of radiant energy by a green pigment called chlorophyll. Absorbed light is converted to chemical energy and this is used to reduce carbon dioxide to carbohydrate. Ultimately the products of photosynthesis are used up in respiration, an exergonic oxidation reaction, and the energy so released is used to power the building and maintenance of biological organization. After participating in metabolism, energy ultimately appears as heat and is irradiated from the surface of the earth in a much more disordered state than the light energy entering the system. Whatever process might go on within the biosphere, therefore, the system as a whole conforms to the second law of thermodynamics since it creates disorder from order.

2.2 The partitioning of energy in organisms

At the level of the individual there is a fundamental distinction between organisms which possess chlorophyll and which can photosynthesize — the autotrophs — and those which lack the green pigment and which cannot manufacture organic products from inorganic precursors. The latter are the heterotrophs. This distinction divides the living world into the Plant and Animal Kingdoms and it also divides the world very crudely into movers and non-movers. Most plants do not need to move to feed because sunlight is all-pervasive, whereas most animals need to move to feed since their food materials are distributed in a patchy fashion over the face of the earth. The need to move for food and to avoid being eaten has brought with it the need to be aware of the environment and this has ultimately led, in man, to the need to reason.

Input, potential energy locked in organic molecules and derived in plants from photosynthesis and in animals from feeding, is partitioned along two major pathways (Fig. 2.2). It may either be used in tissue respiration (R) or in the synthesis of new protoplasm (G). The synthetic process will be considered more fully in the next section.

As already noted, respiratory metabolism results in a large expenditure of energy; larger, in fact, than the activity of the sun. The sun produces 2×10^{-4} watts/g whereas a 70 kg man produces something in the order of 2 watts/g (Broda, 1975). Some of the energy released in metabolism is conserved in the energy-rich, terminal phosphate bonds of molecules like ATP (adenosine triphosphate) and the rest is lost as heat. The whole respiratory process usually requires oxygen but sometimes it can occur under anaerobic conditions. The efficiency of energy transfer from substrate to ATP is less, however, under anaerobic conditions (*ca* 30%) than under aerobic conditions (*ca* 65%) and this is illustrated in Fig. 2.3.

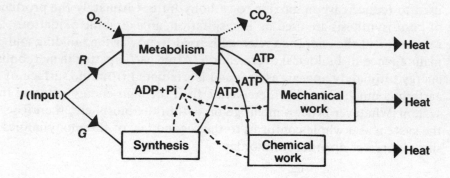

Fig. 2.2 Partitioning of energy within the organism. (With permission from Calow 1977b).

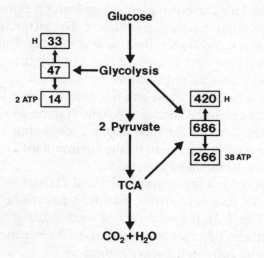

Fig. 2.3 Efficiency of anaerobic glycolysis and the aerobic tricarboxylic acid (TCA) cycle. Figures in the boxes represent the total energy release in kcal (multiply by 4.2 to get the joule equivalent); H signifies that component lost as heat, the rest being used in the formation of ATP.

The ATP serves as a carrier of chemical energy within the cell; it is not the only carrier but it is the main one. In yielding energy it breaks down to form ADP (adenosine diphosphate) and inorganic phosphate (Pi). The energy released in this way is used to power physical work (e.g. contraction), chemical work (e.g. active transport), and the synthetic processes themselves. All these transfers are inefficient and heat is liberated from them.

The ADP and Pi are recycled to the site of tissue respiration (mainly the mitochondria) and are reassociated as ATP. Despite this continuous need for ATP, however, the amount per cell is usually very small (*ca* 0.3–0.002% dry weight) suggesting that the turnover rate of ATP must be extremely rapid. Broda (1975) estimates that 1 g of dry bacterial matter produces 180 g of ATP per day, that ATP in bacteria is turned over 300 000 times per day and that the mean life of an ATP molecule is 1/3 of a second. Even so, the level of ATP in cells remains surprisingly constant over a wide variety of conditions and this indicates that it is under strict control. The same is true of the extent to which the pool of adenosine phosphates is filled with free energy, a property which is conveniently measured by the energy charge of Atkinson (1972):

$$\frac{[ATP] + \frac{1}{2}[ADP]}{[ATP] + [ADP] + [AMP]} \tag{2.1}$$

(where square brackets represent 'concentration of'). Under normal conditions this remains very constant; for example in bacteria the value fluctuates only slightly around a value of 0.85.

Table 2 ATP per dry weight (%) of various stages of the animal life cycle (Data from Zotin, 1972).

Stage	ATP % dry weight
Various cleaving eggs	0.2
Various embryos	0.14–0.18
Adult loach	0.03
Adult mouse	0.0015

At the organismic level it is well known that the intensity of aerobic respiration (R), measured as oxygen inspired, carbon dioxide expired or heat lost, and expressed on a per gram basis, reduces with the size of the whole organism (W). In general, $R = aW^b$ where a and b are constants and $b \doteq 0.75$ (Hemmingsen, 1960; Kleiber, 1961). There is no fully satisfactory explanation of this relationship, though it may be correlated with reductions in respiratory surface, or increases in inert, non-respiratory tissue as the organism gets larger (Calow, 1975). Zotin (1972) attributes it to a general thermodynamic tendency of open systems to approach a steady-state of minimum entropy production (Onsager's reciprocal relation or Prigogine's theorem) but this is not biologically very satisfactory. He also shows that the ATP content of cells reduces with the size of the organism (Table 2). It is likely, however, that this is an effect rather than a cause of reduced metabolism and certainly cellular suspensions extracted from organisms of different size do not show the size-specific reduction in metabolism (Bertalanffy and Pirozynski, 1951; Newell and Pye, 1971). Hence, the size effect must be due to properties associated with the whole organism, not its parts. Precisely what these properties are, however, still remains one of the major puzzles of comparative physiology.

2.3 Problems in trying to assess efficiency

The efficiency of a machine is usually defined as the ratio of its useful output to its input. By analogy it is possible to obtain similar efficiencies for biological systems. Referring to Fig. 2.2, G can be considered equivalent to the useful product of the system and I as the input. Therefore G/I defines the effeciency ratio. However, a number of difficulties arise in trying to apply this seemingly simple idea to real, living organisms. It is necessary in animals, for example, to distinguish between an actual and an apparent input. The food eaten (C) is the apparent input but only a fraction of this (A) is absorbed across the gut wall and is actually available for metabolism; the difference $C-A$ is the energy egested. Also, in 'biological machines' the product, G, is not usually removed from the system but is retained as part of the fabric.

A more profound difficulty arises in connection with the subjective nature of the idea of efficiency itself. In defining efficiency the needs of the subject for whom the job is being done must also be taken into account (Slobodkin, 1962; McClendon, 1975). For example, whereas most people would judge the efficiency of an internal combustion engine in terms of the way it converted potential, chemical energy into the kinetic energy of motion, an eccentric may wish to use the motor to warm his hands. The efficiency of the motor in meeting the needs of the eccentric (E) would be different from the efficiency of the motor in meeting the needs of the normal user ($1-E$). More reasonably, it would be quite plausible to judge the efficiency of a machine in terms of the rate it converted input to output or in terms of its ability to maintain a constant output despite disturbances in the input rather than just as G/I.

Perhaps efficiency is best judged with respect to the function for which a system is designed. This makes the definition more objective by narrowing it down to the needs of a single subject, the designer. Natural selection is the designing force in the biological world and is concerned with the extent to which 'biological machines' produce copies (themselves copyable) of their own genetic instructions. In these terms, therefore, information transmission rather than energy transactions represents the most important determinant of efficiency. However, since as a first approximation the amount of energy put into protoplasm (G) will be related to gamete formation (cf. Chapter 5) it is reasonable to expect that fitness should be roughly proportional to the flow of energy into G. Accordingly, it is possible to decide if 'biological machines' should be designed to maximize G/I, or the speed of production of G from I (e.g. by maximizing the rate of input) or the homeostatic ability of the system to keep G constant despite disturbances in I. Of course all three strategies could be appropriate under different conditions. According to the definition of fitness given above the prime requirement is that the biological systems should obtain energy and produce as many gametes from it in the shortest possible time. This can either be achieved by maximizing I/time for a fixed G/I or by maximizing G/I for a fixed I or by modifying G/I in the face of a variable I. Apart from the very general law, first formulated by Lotka (1922), that the production of G should be maximized, there seem no *a priori* grounds for placing restrictions on how this should be achieved. It is possible, however, that selection would have shifted in the course of time from maximizing speed to maximizing efficiency and homeostatic ability since those niches filled last would be the ones where resources were most difficult to obtain (Calow, 1977a).

2.4 Some estimates of conversion efficiency in heterotrophs

In this section I calculate a 'best possible' conversion efficiency against which actual efficiencies can be compared. Since the metabolic processes are very complex and are likely to vary considerably in quantitative detail from one species to another the results of these calculations must necessarily be very general and very approximate. The aim, however, is not to obtain a precise efficiency but to establish an upper limit to the efficiency of the conversion process and thus to use the reference point in the comparative treatment of data from real systems. Most of what follows in this section is taken from Calow (1977b).

Biomass is made up of polymers and, in heterotrophs, polymers are produced by linking monomers derived from food. For the purpose of this calculation it is possible to express the quantity of linkages per polymer as a mole equivalent and then, knowing the ATP needed in the formation of the linkages, to progress to an estimate of energy required in the building of polymers from monomers. One dry gram of 'average' metazoan biomass contains approximately 0.01 mol of polymeric linkages

Table 3 Average number of ATP's and efficiency of energy yield from ATP in the formation of polymeric bonds. (After Calow, 1977b).

Polymeric bond	Efficiency of energy yield from ATP	No. of ATPs per bond	Proportion of bonds in biomass
Glycosidic	80	2	12
Ester	34	8	6
Peptide	20	3	82
Weighted average	28	3	–

(Morowitz, 1968), and has a heat of combustion (ΔH) of 23.1 kJ (5.5 kcal) (Cummins and Wuychek, 1971). On average, three phosphate bonds are hydrolysed in the synthesis of each polymeric linkage (Table 3) and the free energy (ΔG) in going from 3 ATP to 3 ADP is approximately 86 kJ/mol (21 kcal/mol). Hence (0.01 × 86) kJ or 860 J are required to synthesize 1 g of biomass. If we assume that we are dealing with an aerobic system with tissue respiration operating at 65% efficiency (Krebs and Kornberg, 1957) then (860 ÷ 0.65) or 1323 J would be required to generate the necessary ATP. Hence 463 J would be lost as heat and the total amount of energy required in the synthesis of 1 g of biomass would be 23 100 J to be put into the biomass itself, plus 1323 J to generate the ATP, or a total of 24 423 J. The average efficiency of transfer of energy from ATP to the polymeric linkages is 28% (Table 3) so that a further (860 × 0.72) J or 619 J would be lost as heat.

The net conversion efficiency is therefore:

$$(1 - \frac{\text{cost of production}}{\text{total amount used}}) \times \frac{100}{1} \qquad (2.2)$$

$$= (1 - \frac{1082}{24\ 423}) \times \frac{100}{1} \simeq 96\%$$

This result shows that potentially, biological systems can be extremely efficient in converting raw materials to products, far more efficient in fact than any technological system. Actually, the above result is an absolute best efficiency since it must still be reduced by the cost of assembling the super-molecular organization of biological systems and by the cost of maintaining the biomass once it has been produced. Also, in the higher heterotrophs, the gross conversion efficiency will be reduced with respect to the net conversion by an amount equivalent to the inefficiency of the digestive process. Unfortunately these extra costs can only be computed roughly since they are not physico-chemical requisites of the conversion process but depend rather on the biological structure and function of the system. As such they may show considerable variation from species to species and from tissue to tissue.

The cost of building a super-molecular organization out of polymeric constituents would appear to be negligible. The heat of combustion of the organized cellular organelles and the cells themselves is very nearly equal to the summed heats of combustion of the constituent macromolecules (e.g. Meyerhof, 1924; Penning De Vries *et al.*, 1974). However, the cost of maintaining the biomass once it has been produced may be high. It involves the cost of turnover of the synthesized molecules, the cost of transport of molecules into and within cells, and the cost of mechanical work.

Turnover of materials within cells, even bacterial cells (Section 4.2), may be considerable. The cost of turnover is attributable to the work involved in synthesizing the replacement polymers and to the energy lost during the breakdown and excretion of the replaced molecules. Though there is now considerable information on rates of loss, particularly for proteins, little is known about the extent to which the breakdown products are re-cycled. Furthermore, it is difficult to put any general figure on turnover, since different molecular species have significantly different half-lives. Some proteins for example, may have half-lives measured in minutes whereas DNA has a life equivalent to that of the cell in which it occurs (Section 4.2). A minimum estimate, therefore, would be that, on average, each molecule turns over once in the life of the system and that monomers are re-cycled. To allow for this twofold turnover, it is necessary to raise the metabolic cost of 1082 J by a factor of about two, i.e. to 2164 J and reduce the conversion efficiency to $[1 - [2164/(24\ 423 + 1082)]] \ 100$, or 92%.

Estimates on other aspects of the maintenance cost are even less precise. The cost of molecular transport, for example, undoubtedly varies from system to system and from time to time. Experimental findings with ouabain, a drug which blocks active transport, have shown that in nervous tissue (Gonda and Quastel, 1962), between 30 and 50% of basal metabolism may be accounted for in terms of the transport system, and similar results have been obtained for other tissues (Maizels, Remongton and Truscoe, 1958; Paul, 1965). Hence if the figure of 50% were applied generally it would raise basal metabolism by another factor of two from 2164 J to 4328 J. Since the site of synthesis of a molecule may not be that at which it will be used, energy is also required for molecular transport within the cell. ATP is similarly required for the transport of molecules

Table 4 Conversion efficiencies for several groups of heterotrophs. (Data from Calow, 1977b).

	k_1	k_2
Bacteria	*	58
Protozoa	*	50
Moulds, yeast	*	60
Eukaryotic cells in culture	*	60
Embryos	*	60–70
Young post-natal poikilotherms	50	50–80
Young post-natal homeotherms	35	50–70

k_1 = (G/energy ingested to alimentary system) 100
k_2 = (G/energy absorbed into metabolically active site) 100
* in most cases there is no defaecation, hence k_1 is inappropriate.

from where they are used to the lysosomes where they are ultimately broken down (Natori, 1975). However, Penning De Vries *et al.* (1974) have calculated that adjustments for this kind of process may make as little as a 1% difference to the estimate of overall efficiency. Therefore, adjustment of the conversion efficiency for the total, potential transport cost gives

$$\left(1 - \frac{4328}{25\,505 + 2164}\right) - 0.01,$$

or approximately 84%.

As with active transport, the cost of mechanical work varies from system to system. In growing systems, however, most cells will at some time spend energy in the mechanical separation of nucleic acid strands and daughter cells during mitosis and both theoretical (Frey-Wyssling, 1953) and experimental (e.g. Marsland *et al.*, 1953) estimates show the cost to be significant. Though there is little rise in the respiratory rate of cells during cleavage (Zeuthen, 1950a, b and c) it is likely that most of the metabolism at this time is concerned with mitotic activity. Furthermore, mitosis is apparently carried through by a store of ATP accumulated before the onset of division (Bullough, 1952; Giese, 1973). The cell and nuclear membranes may also undergo contraction and relaxation from time to time and organelles, like mitochondria, undergo swelling, contraction and movement cycles. It is extremely difficult to put any estimate on the cost of these processes but it would seem not unreasonable to assume that they are likely to reduce the overall efficiency of the system below 80%. Hence the best possible efficiency that can be expected from any growing heterotrophic system is likely to lie somewhere between 70 and 80%. The very best possible efficiency, excluding the cost of maintenance, is between 80 and 95%.

After calculating the theoretical maximum it is now possible to compare it with actual efficiencies obtained from experimental data. Table 4 contains a review of the ranges of efficiency found in various heterotrophic systems and shows that the theoretical maximum is not achieved by all systems; as expected (Section 2.3) selection has not automatically favoured the maximum G/I. Young poikilotherms most nearly approach the theorectical expectation, whereas homeotherms have probably sacrificed some efficiency for the advantages of increased homeotaxis (e.g. in body temperature). On average, unicellular organisms have lower efficiencies than metazoa, and here the evolutionary emphasis may have been on speed of turnover rather than efficiency. The figures for the metazoan cells are probably underestimated since cellular activity is probably greater in culture than in the tissues *in vivo*. Finally, the gross efficiencies of the metazoa and less than the net efficiencies, since digestion and absorphan are themselves inefficient. This will be discussed further in Section 2.6.

2.5 Photosynthetic efficiency

Photosynthetic efficiency is the ratio of biomass produced to radiant, input energy. However, complications arise since input may either be measured and expressed in terms of the total amount of light energy reaching the plant or only as that absorbed by the tissue. Similarly, production may be expressed as the gross amount of material resulting from the photosynthetic process or as the net amount after the subtraction of respiration. In the field, efficiencies for macrophytes based on net production, are usually less than 5% (Phillipson, 1966), though in carefully managed plots may rise to between 12 and 20% (Wassink *et al.*, 1964). In laboratory algal cultures, however, efficiences of 20–30% can be achieved under the appropriate conditions (Burlew 1964).

Since there is still some uncertainty about the chemical basis of photosynthesis it is not yet practicable to calculate what efficiencies might be achieved by this process under ideal conditions. It can be calculated, however, that the efficiency by which the products of photosynthesis are converted to biomass is in the order of 60–70% (Penning De Vries *et al.*,

1974). This is slightly lower than the efficiency by which animals convert absorbed material to protoplasm calculated in the last section because prior to the synthesis of plant protoplasm some of the carbohydrates must be transformed to fatty acids and amino acids.

2.6 Metabolic adaptation

This section is concerned with establishing, more precisely, the effect on fitness of diverting input energy into new protoplasm. The argument follows Alexander (1967) and applies mainly to animals. Consider the following energy budget:

$$C - F = A = Pg + Pr + H + Ex \qquad (2.3)$$

where: C = food consumed; F = food defaecated; A = food absorbed across the gut wall; Pg = energy used in growth; Pr = energy used in gamete production; H = total heat loss; Ex = energy loss as excreta. Pg and Pr can be lumped together as protoplasm produced (G). G/C is the gross efficiency sometimes referred to as k_1; G/A ($\equiv G/I$ in last section) is the net efficiency, sometimes referred to as k_2. In practice it is often difficult to measure faeces and excreta separately so they are sometimes combined as 'rejecta' ($rej = F + Ex$). The term $(C - rej)/C$ defines that proportion of ingested energy which is available for metabolism and synthesis (call this u). H, the heat loss, will consist of a component associated with basal metabolism and a component arising from the synthetic activity. Since the latter component will be related to the intensity of synthesis it is useful to express it as a function, g, of G. Hence the energy budget reduces to:

$$C.u = g(G) + Act \qquad (2.4)$$

where Act = the component of H due to activity. Alexander (1967) has solved the equation for teleost fish. Here $u \simeq 0.8$, $g \simeq 2$ and G is about 10% of Act. For convenience set C to 100; then:

$$100 \times 0.8 = 2(6) + 68 \qquad (2.5)$$

and it can readily be seen that a small change in C (say 100 to 101 or a 1% increase) would have a larger proportional effect on G (6 to 6.4 or a 7% increase). Similarly, a 1% decrease in Act should lead to a 3% increase in G. Assuming that changes in G have a direct and proportional effect on gamete production (cf. Section 5.1), it is possible using equation 2.4 to calculate the coefficient of selection of genotypes likely to produce these effects (see Chapter 1).

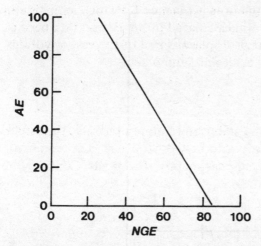

Fig. 2.4 Relationship between absorption efficiency (*AE*) and net growth efficiency (*NGE*) in a selection of aquatic poikilotherms. (Data from Welch, 1968).

If a fish of one genotype can obtain 1% more food than that of another and from this extra ration it can produce 7% more eggs then the coefficient of selection of the former with respect to the latter is 0.035. Again assuming no genetic dominance (Chapter 1), then the gene for eating more will increase relative to the 'wild type' by 2.7 times every 2/0.035, or 57 generations. If the original frequency of the gene in the population was 1% then it will rise to 2.7% in 57 generations and 99% in 2109 generations. This may take 100 000 years but such a time interval is short in the 100 million years which have been available for fish evolution (Alexander, 1967).

Though based on a number of simplifying assumptions (Priede, 1977), the above calculation does illustrate how small alterations in the energy budget enabling more energy to be accumulated in *G* can have a profound effect on fitness. It is reasonable to expect, therefore, that the metabolism of an organism will be finely adapted to the conditions, both biological and physical, in which it operates. Many examples might be chosen to illustrate this point (Calow, 1977a) but I shall only consider two; the involvement of digestive efficiency in stabilizing the growth efficiency of herbivores and carnivores and the effect of mucus production on the metabolism of triclads.

Fig. 2.4 is taken from a review by Welch (1968) and summarizes data from 14 species of cold-blooded, aquatic animals. There is a clear negative

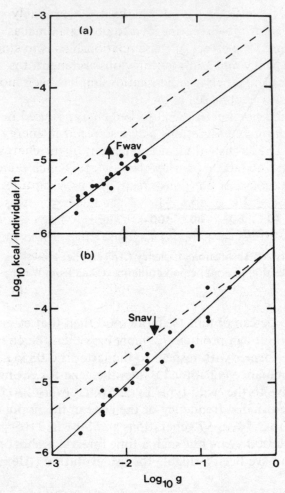

Fig. 2.5 The relationship between the logarithm of respiratory rate (kcal/h; multiply by 1.2 to obtain watts) and the logarithm of body size (gram, shell-free wet weight) in flatworms (a) and snails (b). The broken line is Hemmingsen's standard regression line for poikilotherms (Hemmingsen 1960). Fwav = average sized flatworms. Snav = average sized snails. The respiration of the 'average animals' in each case is significantly lower than that expected from the standard line. The difference is constant throughout the size range of flatworms but the 'snail line' converges to the standard line as the snails get bigger. (With permission from Calow 1977a, *Adv. Ecol. Res.* **10**. Copyright by Academic Press Inc. (London) Ltd.)

relationship between net growth efficiency and absorption efficiency (food absorbed divided by food ingested). In general, the high absorption efficiencies were associated with the meat-eating habit and low efficiencies with a vegetarian diet. Similarly, low net growth efficiencies were associated

with carnivory and high net efficiencies with herbivory – probably
reflecting the relative cost of movement needed to obtain animal as
opposed to vegetable food. The effect of these relationships is to stand-
ardize gross growth efficiency and from our previous argument it is
reasonable to expect that this finely geared relationship has been moulded,
at least in part, by a positive act of selection.

Not all the non-respired portion of the absorbed energy is used in growth
and reproduction; excretion and secretions may also result in energy loss.
Usually the latter represent an insignificant component of the energy
budget but it has been estimated that freshwater triclads, which secrete
copious mucus for the purpose of both movement and prey capture, may
lose as much as between 20 and 70% of their ingested energy in this way
(Teal, 1957; Calow and Woollhead, 1977a). In this group, however,
compensatory economies seem to be made through reductions in the cost
of maintenance. Fig. 2.5a, for example, compares the relationship
between respiratory rate and size in several species of triclad with the
standard line obtained for other metazoan poikilotherms under approximately the same conditions. Though the slope of the 'flatworm line' does
not differ from the slope of the 'standard line' there is a clear difference
(which is statistically significant, Calow, 1977a) between the level of
these lines. On average, flatworms respire 2.75 (0.440 log decades) times
less than most other metazoa of the same size: that is, they have a lower
cost of maintenance. A similar device seems to have evolved in the snails
which also produce copious mucus (Fig. 2.5b), and food-limited, top
carnivores like spiders (Anderson, 1970).

2.6 Metabolic adaptability

It is important, as we shall see later (Chapter 5), for organisms to reach
prescribed stages in their development at particular times. Those organisms
which can do this despite environmental disturbance must be at an
advantage over less adaptable contemporaries. Hence, metabolic adaptability, or homeostatic ability, is likely to be an important adaptation.

A common form of disturbance in animals occurs through reductions
in food supply. That this does elicit metabolic compensation is amply
demonstrated by the fact that growth efficiencies often increase with
reducing ration (see review, Calow, 1977b), at least up to some optimum
(Elliott, 1975), that growth rate is not directly proportional to food
supply (Calow, 1973) and that growth 'spurts' and 'tracking' phenomena
often occur after a period of time when growth has been partially or

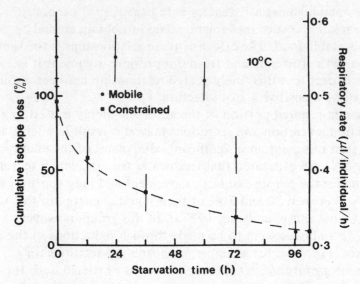

Fig. 2.6 Effect of starvation on the respiration of constrained and mobile
P. contortus. The broken line represents the time-course of gut emptying
and this is well-correlated with 'constrained metabolism'. (With permission
from Calow 1974).

completely suspended (Tanner, 1963). In animals, similar adjustments
may be observed after reductions in the quality rather than the quantity
of the ingested food.

Behind these whole-organism responses there are physiological and
behavioural changes. During complete famine, for example, behavioural
and physiological strategies are deployed to increase food-finding ability
and to conserve energy. Often these requirements conflict and it is
necessary to compromise between them. Fig. 2.6, for example, illustrates
the respiratory response of *Planorbis contortus* (a freshwater pulmonate
snail) to starvation. The rates of mobile and constrained (immobile)
individuals are plotted separately against starvation time. Starving snails
move more than fed snails, presumably as a searching response. Respiratory
rate in these individuals is therefore maintained at a high level until
movement is finally reduced in weakened individuals. Searching is
expensive in energy but the cost is offset in *P. contortus* by reductions
in the resting metabolic rate as illustrated in the immobile snails.
Whether or not a species invests energy in searching will depend on the
cost-effectiveness of the extra effort. In general, species feeding on
mobile food become quiescent upon starvation because there is a good
chance that the food will come to them, whereas species which feed on

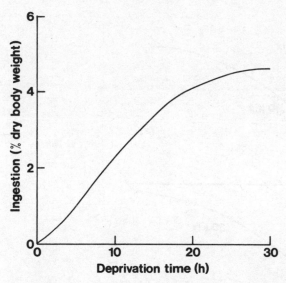

Fig. 2.7 Relation of food deprivation time to appetite (voluntary food ingested as % body, dry weight) in fingerling, Sockeye Salmon. (After Brett, 1971).

immobile food need to invest energy in searching since they must go to their food (Calow, 1977a).

When a partially starved individual ultimately finds food again it tends to eat more (Fig. 2.7) and digest it more efficiently (Calow, 1977a). The relationship between ingestion rate, absorption efficiency and starvation time is not simple but the effect is clearly to compensate, at least in part, for what has been lost in terms of production over the intervening period.

Similar adaptability may be observed in the response of plants to reducing light levels. Differential growth causes the leaves of many terrestrial macrophytes to 'bend' away from the shade towards the sun. Furthermore, plants grown in low light intensity are usually more efficient at photosynthesis than plants grown at high intensities. This is well-illustrated by some results obtained on *Chlorella vulgaris,* a green, unicellular alga, by Steemann Nielsen *et al.* (1962) (Fig. 2.8). *Chlorella* was grown at various light intensities in dilute culture so that there was no appreciable shading. Two sorts of culture were used; one conditioned to high light intensity, the other conditioned to low light intensity. The curves relating photosynthesis to light intensity were quite different for each kind of culture; the 'low light cells' being more efficient than the 'high light cells' at low light intensities. The greater efficiency of the

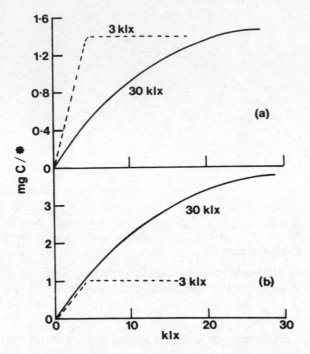

Fig. 2.8 Photosynthetic rate (mg C) of *Chiorella* conditioned to a light
intensity of 3 klx (broken line) and 30 klx (unbroken line) as related to
light intensity. * = 10^9 cells in (a) and mg chlorophyll in (b). (After
Steemann Nielson *et al.*, 1962).

'low light cells' could be accounted for in terms of them having a higher
concentration of chlorophyll because when photosynthesis per unit
chlorophyll was plotted against light intensity there was no difference
between the cultures at low light intensity. This response corresponds to
the production of 'shade' and 'sun' leaves by some higher plants (Boardman,
1977).

3 Life as a cellular system

The organism as an open system takes in and gives up matter and energy. At the cellular level growth occurs when the material and energy in excess of other metabolic demands are used to form more cells, to fill out existing cells and to produce extracellular materials.

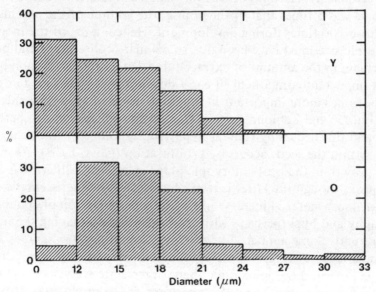

Fig. 3.1 Proportional distribution of cells taken from the bodies of young (Y) and old (O) triclads. The species was *Dugesia lugubris*. Cells were obtained by mincing tissue in a solution of 0.5% E.D.T.A.

Fig. 3.1 shows how the relative proportion of different sized cells changes in young (just hatched) and adult (full grown) triclads. The old organism contains cells of larger average size than the young organism indicating that the increase in organismic size may be explained, at least in

37

Life Cycles

Table 5 Changes in cellular and subcellular composition of liver in the male rat during post-natal growth. The data show that organ growth involves a combination of cellular hypertrophy and hyperplasia. Furthermore, increases in cell size are associated with increases in DNA, suggesting an increase in ploidy. (Data from Fukuda and Sibatani, 1953).

Body weight (g)	Liver weight (g)	Age (days)	Cell no.	Cell mass (ng)	DNA per cell (pg)
12	0.30	10	168	1.79	5.90
25	0.98	21	445	2.20	5.09
50	2.60	31	668	3.90	9.30
100	5.70	41	1060	5.36	11.10
200	8.10	80	1270	6.37	11.10
340	12.00	182	1790	6.70	11.40

part, by cellular hypertrophy. However, whereas the cell volume of the adult worm was 5 times that of the young, the volume of the organism as a whole rose 100 times during development. Hence, most of the increase in organismic size must have been due to cellular proliferation and possibly to an increase in the amount of extracellular material. Cell division is also the most important component of early development in the rat but cell growth becomes more important in the later stages of growth in this species (Enesco and Leblond, 1962). Plant fruits increase in size before ripening and this occurs by cellular hypertrophy, but growth of the embryo, within the seed, occurs by proliferation (Goss, 1973). As in animals, growth in the post-embryonic plant is by both cell growth and proliferation, though the latter is restricted to specific meristematic sites. Therefore, most metazoa increase in size by a combination of cellular hypertrophy and hyperplasia — with the latter usually playing the most important part. Some animals, like rotifers, nematodes, tardigrades and appendicularians, however, have a constant number of cells from a very early stage in their development and here organismic growth occurs mainly by cellular hypertrophy. The cells of rotifers, for example, may grow to sizes which approximate to the whole size of the freshly-hatched larvae.

It is often assumed that cellular hypertrophy occurs mainly by the proliferation of the cytoplasm. In some cases, however, increases in the size of individual cells are accompanied by an increase in the amount of chromosomal material; that is, by the development of polyploidy. Table 5 presents data taken from Fukuda and Sibatani (1953) for the rat liver and illustrates how the increase in mass of just one organ may involve a complex series of cellular events. During the first 20 days, cell proliferation is important; from day 20 to day 40, changes in cell size are accompanied by increases in amount of DNA per cell; from day 40 onwards, increases in cell size occur without changes in DNA.

Of course changes in cell standing crop need not reflect the rate at which cells turn over within the organism. The number of cells present at any one time is a balance between proliferation and loss. When proliferation exceeds loss, the size of the cell population rises; when cell loss exceeds proliferation, population size falls. It is even possible for there to be a turnover of cells in a steady-state population, provided proliferation just balances loss. Usually metazoan organisms are made up of a number of sub-populations of cells each showing different proliferative properties. In some tissues, cells have permanently ceased to divide (like striated muscle cells and nerve cells in vertebrates and supporting and transporting tissues in higher plants); in others, cells divide only under exceptional circumstances (e.g. after wounding); and in yet others, cells divide all the time (e.g. cells in the skin and gut epithelium of vertebrates and in the terminal meristems of plant roots and shoots).

3.1 The cell cycle and growth

Within the organism, cell division consists of a duplication of genetic information and then a separation of these duplicates into portions of cytoplasm which ultimately break apart. This is mitosis. The cellular products of mitosis, though containing the same gene compliment as the parent cell, must be smaller. In closed, non-feeding, embryonic systems this means that cell size reduces with each successive cleavage (Fig. 3.2) but in feeding organisms cell size is usually reinstated between divisions. Also, before the cell can divide again, the genetic material (DNA) must be replicated and this process is restricted to a specific period in the interphase (period between divisions) sandwiched between two periods when there is no synthesis of nucleic acid. Cells therefore cycle from mitosis (the M phase) through a gap (the G_1 phase) to the synthesis of replicate DNA (the S phase) and then through another gap (the G_2 phase) back to mitosis (see Fig. 3.3). Cell growth, unlike nucleic acid synthesis, occurs throughout interphase and may take many forms (see Fig. 3.4). The details of this cell cycle may vary, particularly in timing, from one tissue to another but in general form the cycle is very constant throughout all tissues be they animal or plant. In some tissues, of course, cells pass out of the mitotic cycle into a non-dividing state. This may occur irreversibly, as in nervous tissue, or reversibly, as in liver. For example,

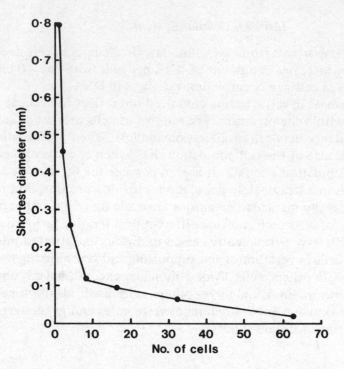

Fig. 3.2 The relationship between cell size and number in the early developmental stages of the Clawed Toad, *Xenopus laevis*.

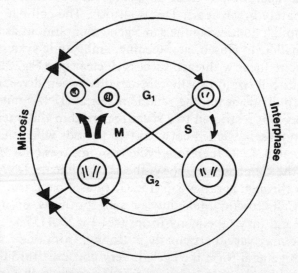

Fig. 3.3 The cell cycle. This is elaborated further in Fig. 3.6, where the relative positioning of resting phases is illustrated. See text for further explanation.

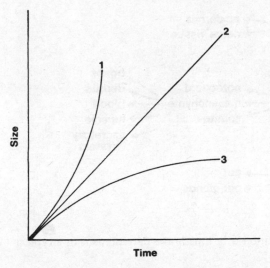

Fig. 3.4 Possible types of cell growth. Curve 1 has been described for yeast (Mitchison, 1958), curve 2 has been described for *Escherichia coli* (Schaecter *et al.,* 1962) and curve 3 for *Amoeba proteus* (Prescott, 1955). Other cells, including those of metazoa, also fit these curves (Sinclair and Ross, 1969).

in the latter case mitotic activity is reinstated after wounding or hepatec-tomy. It has been suggested that this resting phase be represented as a separate stage, referred to as G_0, in the cell cycle, into which cells feed after mitosis (Baserga, 1976). Some cells also enter a resting state from G_2 but this is more rare. These points are illustrated later in Fig. 3.6.

Cell size was once thought to be important in controlling cell division — cells were thought to have to grow to a critical size before they could divide. Experimentally, however, it has been possible to inhibit cell growth without stopping division and it now seems clear that the size of cells is only one of the parameters involved in the control of proliferation. The quality of the internal environment in terms of nucleic acids, energy charge and protein supply may also be important. We shall return to this question again in Section 3.3.

3.2 Differentiation and growth

Cell division results in the construction of a multicellular organism from a single-celled zygote and together with cell growth causes the organism to get bigger. As well as coming to contain more cells, the growing

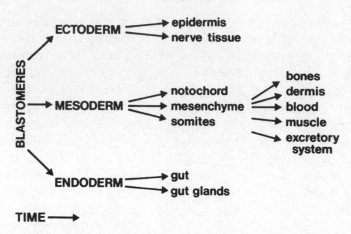

Fig. 3.5 Increased differentiation with time in the development of a typical vertebrate.

organism also comes to contain more types of cell. That is to say, different cell lines begin to make different sorts of molecules and to take on different metabolic properties. This process is known as differentiation. On a phylogenetic basis we have already seen that bigger organims are more differentiated than smaller organisms (Table 1) and in a 'Haeckelian' fashion this is also true of ontogenetic increases in size (Fig. 3.5).

The products of mitosis each contain replicas of the zygotic genome. Hence, in order that brother and sister cells can take on different metabolic roles within the organism there must be a differential expression of the common genetic instruction. This seems to involve the differential activation of an inert genome rather than the inactivation of an active genome. Germ cells, for example, do not produce the same range of molecules needed by the adult, otherwise mammalian zygotes would be bright red due to the present of haemoglobin. Furthermore the tightly coiled nature of the eukaryote DNA (Section 1.6) puts the information on the inside of the molecule and means that it is not immediately available for transcription. The DNA must 'uncoil' prior to read-off and, as previously suggested, 'puffing' seen in the giant chromosomes of certain Diptera is thought to be the visible manifestation of this process. Certainly, work with tritiated uridine has shown the puffs to be localized sites of intense RNA synthesis (Ficq and Pavan, 1957).

There are, of course, several good reasons for the genetic instructions being called into play when required, rather than being progressively

switched off as development proceeds. First, holding the genetic information in latent form protects it to some extent against thermal noise;
second, time and energy are not being wasted in manufacturing surplus
molecules; finally, the alternative would mean that the zygote and early
blastomeres would contain a chaotic assemblage of every conceivable
molecule needed by the adult. Control at transcription seems to be the
most sensible strategy and does indeed appear to be the most common
form of regulation. However, in eukaryotes there is also some evidence for
the control of gene expression at the level of translation since here mRNA
is more stable and long-lived than in bacteria (Watson, 1970; Ford, 1976).
Furthermore, in the water mould, *Blastocladiella emersonii*, zoospore
germination, which involves dramatic changes in cell architecture, may go
on independently of protein synthesis (Soll and Sonneborn, 1971),
suggesting that differential protein synthesis is not an exclusive and
sufficient explanation of phenotypic change.

The molecular basis of transcriptional control in metazoa has not so
far been elucidated, though because of the remarkable evolutionary
constancy of genetic coding and protein synthesis it is not unreasonable
to expect it to resemble situations in bacteria where appropriate parts of
genome are brought into play under appropriate conditions; for example
the response of *E. coli* to lactose (Chapter 1). As already noted, however,
in eukaryotes extra regulation is likely to be exerted on the translation of
mRNA. In any event the most straightforward models for the control of
differentiation involve signals which emanate from outside the nucleus.
In embryos which show mosaic development (e.g. annelids and molluscs)
the developmental fate of every cell is rigidly determined and controlling
signals presumably arise from within each cell. Alternatively, in embryos
which show regulative development (e.g. sea urchin) it can be demonstrated
experimentally that individual cells may follow one of several developmental pathways. Here, developmental fate depends upon cellular interaction and presumably on signals which arise outside individual cells. In
keeping with the 'central dogma' these signals must act as switches, bringing into action particular parts of the genome, and not as inputs of
detailed information. Furthermore, the signals must operate in such a way
as to produce a specific pattern of differentiation because different cell
types are organized in a definite, not a haphazard, way within the
organism. There seem to be two ways by which this might occur. First,
how a cell differentiates may depend on the differential distribution of a
single, signalling 'substance' (e.g. as a gradient) and here pattern formation would depend on how the intensity of the signal changed from one

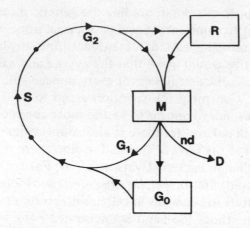

Fig. 3.6 The cell cycle. The basic cycle is from mitosis (M) through G_1 to
synthesis (S), through G_2 back to M etc. Some cells differentiate, enter a
non-mitotic state (nd) and ultimately die (D). The time spent in nd varies
from tissue to tissue. Some cells may pass out of the cycle into a resting
state from which they ultimately pass back into the cycle. Stem cells from
vertebrate liver, renal tubules and the salivary glands enter a G_0 resting
state immediately after mitosis. More rarely, some stem cells, e.g. those from
Hydra and triclads, enter a resting state (R) from G_2 and immediately prior
to mitosis. These latter are in a state of continuous readiness for division.

part of the embryo to another. In this way each cell might identify its
position relative to the intensity of the signal and differentiate accordingly.
Alternatively, how a cell differentiates may depend on the type of signal
it contacts and here pattern formation would depend on qualitative
differences in many signalling substances throughout the embryo. The
simplicity of the one signal idea, which has been advanced most consist-
ently by Wolpert (1969, 1970), gives it appeal, but at the same time, it
is difficult to imagine how, out of so much simplicity, there could arise
so much complexity in the fully-formed organism. In consequence,
several plausible qualitative theories have been advanced (Horder, 1976).
So far, no one experiment has completely negated either the qualitative
or quantitative hypothesis and the exact nature of the signals remains
something of a mystery. It seems likely, though, that the signalling substances are small, non-specific molecules (Horder, 1976).

Whatever the mechanism involved in controlling differentiation it is
now well established that as cells become more specialized they lose their
capacity for division (Bullough, 1967). Switching on specialized parts of
the genome leads to a repression of that part involved with mitosis.

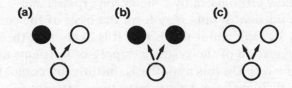

Fig. 3.7 The fate of the products of cell division. Cells represented by open circles are capable of further division, cells represented by closed circles are not.

Cells emerging from mitosis either make ready for mitosis by passing around the cell cycle or they specialize and pass out of the cycle as illustrated in Fig. 3.6. Once on this course, reversion to a mitotic state is unusual but can occur, particularly after a gross disturbance of the tissue Section 3.1. Usually, however, functional cells begin to age and ultimately die. In tissues with a high turnover (e.g. epidermis), cell life-span may be very short, a matter of hours, whereas in those tissues like the nervous system, which cease to divide shortly after embryogenesis, the life-span of the specialist cells may be as great as the life of the whole organism. Of course, both products of mitosis need not follow the same course and there are, in fact, three possibilities (Fig. 3.7). One cell may remain mitotically active while the other specializes (a); both may specialize (b) or both may remain proliferative (c). The first method is probably typical of tissues with a continuous turnover (e.g. epidermis and subcutaneous glands, Ebling, 1955; Leblond and Cheng, 1976); the second probably occurs in the later stages of development in non-dividing tissues and the third must occur in the early stages of development.

3.3 Nutrient supply and cell division

In expanding cell populations, matter and potential energy are required for the growth of the products of mitosis. Under culture conditions, the conversion of nutrients to cellular protoplasm occurs with an efficiency of around 60% (Table 4) and *in vivo* it is likely that this reaches the maximum, theoretical level of 70–80% since here the cost of movement is likely to be negligible (Calow, 1977b). Certainly, embryos with a large proportion of proliferative tissue do reach the theoretical limit of efficiency in the later stages of their development (Table 4). Therefore, if the cellular conversion process occurs at a constant, maximum efficiency it may be postulated that cell systems depend only on a sufficient supply of energy and matter to produce new protoplasm by cellular hyperplasia and hypertrophy. Therefore nutrient supply may form the basis of the control of cell turnover (Calow, 1977b); and if this is true it is likely that the control is effected by regulation of the systemic supply of nutrients and of their transport into the cell. On this hypothesis, mitotically competent cells are prevented from dividing by nutrient restriction. Alternatively, differentiated cells do not divide because they no longer have the genetic competence and therefore the necessary molecular machinery.

One apparent anomaly remains: if cellular yield is substrate-dependent and occurs with a constant efficiency how, as described in Section 2.7, does the conversion involved in organismic growth occur with a variable degree of efficiency? Metazoa consist of two main cellular compartments: one specialized for metabolism (i.e. physical and mechanical work) but not mitosis, the other specialized for mitosis but not metabolism. Since it is proposed that the latter grows at a rate directly proportional to nutrient supply the overall efficiency of the system will depend on the partitioning of input energy between the compartments. If this partitioning is sensitive to external substrate supply then growth rate may be maintained, despite reductions in this supply, by diverting more energy to the mitotic compartment and less to metabolism. This would cause the overall conversion efficiency of the system to rise with reducing nutrient levels as is observed. This general idea of metabolic adaptability has already been discussed in Section 2.7.

4 Life as a dynamic steady-state

According to the thesis set forth in Part 1, organized living systems are continuously subjected to the entropic forces of disorganization. If the organized whole is to persist, the disorganized part must be removed and replaced. Therefore, living systems are inevitably in a state of flux. There has been some argument as to exactly how dynamic the body state is (Thompson and Ballou, 1956), but since the pioneering work of Schoenheimer (1946) biologists have never lost sight of the dynamic view of life.

4.1 Tissues in flux

One of the main reasons for our increasing awareness of the dynamic flux behind the unchanging external appearance of the organism has been the increasing availability and use of radio-isotopes in research. The isotopes of an element differ in physical properties from each other but this rarely affects their chemical and biochemical properties. In consequence, they may be used to label specific compounds within the organism and thus to measure their loss from the body. Radio-isotopes are particularly useful because they can be measured quite easily, even at low concentrations, using appropriate counting techniques.

Table 6 presents the half-lives of several tissues of the rat and was obtained from the long-term, tritium labelling studies carried out by Thompson and Ballou (1956). After the labelling period (> 6 months) the loss-rate of tissues was determined from the rate of disappearance of label. All the tissues turned over, but some more rapidly than others and for each there appeared to be two major components — one with rapid turnover, the other with slow turnover. It is difficult to be very precise, though, about turnover rates from these data since some of the labelled,

Life Cycles

Table 6 Half-lives (days) of various body constituents of the rat, measured by a tritium labelling technique. In each tissue there were two major components, one long-lived, the other short-lived. (Data from Thompson and Ballou, 1956).

Tissue	Long-lived component (*L*)	Short-lived component (*S*)	Relative masses *L/S* x 100
Carcass	130	22	47
Liver	140	12	3
Lung	320	10	14
Kidney	180	8	11
Stomach	300	20	20
Small intestine	160	17	9
Large intestine	180	30	13
Brain	150	16	54
Pelt	110	11	67
Muscle	100	16	40
Bone	240	16	72

breakdown products may have been re-cycled before release. As a result, turnover was probably underestimated. On a comparative basis, however, it is obvious from the data that surface tissues, which are liable to damage, have a more rapid turnover than deep tissues (like liver and kidney) which are not subject to either physical or chemical disturbance. Once more, this underlines the fact that one of the most important aids to survival, in any long-lived organism, is its ability to maintain tissue turnover and thus to repair damage by replacement.

What Thompson and Ballou were measuring within each tissue was the loss of tritium. This may have been going on through the loss of intra-cellular material or by cellular breakdown itself. In either case, however, what was actually being lost was tritiated molecules. Molecular turnover is the basis of the dynamic steady-state and so it is this process we consider first.

4.2 Molecules in flux

The rate of molecular turnover may vary considerably from compound to compound. Some proteins have half-lives in the order of minutes whereas DNA and collagen molecules may last as long as the organism itself. Molecules of the same type may have different half-lives in different parts of the same organism and different 'species' of the same compound may differ greatly in turnover rate even within the same organ (Table 7).

Table 7 Half-lives (days) of various protein fractions in the whole rat (data from Borsook, 1950) and in one organ, the liver of a rat (data from Schimke, 1975).

Whole rat	
Total protein	17
Plasma proteins	6–7
Carcass proteins	21
Liver	
Ornithine decarboxylase	11 min
Catalase	1.4
Glucokinase	1.3
Lactate dehydrogenase	16

As a general rule, structural and informational molecules have greatest stability whereas those substances, like enzymes, which play an active part in metabolism turn over very rapidly. This is, of course, to be expected. Because of their activity, 'working molecules' are most susceptible to damage and denaturation (Section 1.6). At the same time, turnover itself is a risky process since mistakes may easily occur in transcription, translation and synthesis. Those molecules which carry errors or which have been damaged, are replaced according to the instructions carried on protected inert 'genetic molecules'. It is essential, therefore, that the informational molecules are protected against damage and this inevitably means that they cannot be allowed to do very much or turn over very rapidly. Limited repair is possible, however. For example, a constant succession of monitoring endonucleases that are coded to recognize certain abnormalities move around the genome. Once damage is recognized the endonuclease makes an incision on the 5' side of the lesion and an exonuclease then removes the offending strand. Next, polymerases insert the appropriate nucleotides in the gap and the whole new strand is sealed into the space by ligase (Marx, 1973).

The mechanism behind molecular turnover is best understood for the proteins (see collection of papers in Schinke and Katanuma, 1973). Degradation proceeds under the action of specific protease enzymes. It is a sequential process first involving the cleavage of the large protein molecules to smaller polypeptides and then the breakdown of these into amino acids. Specific enzymes seem to be involved at each stage but they are not likely to be molecule-specific otherwise each enzyme would itself require a protease and so on. Degradation also requires energy in the form of ATP, and this is probably needed for the active transport of proteins into the lysosomes — subcellular organelles which are the actual sites of breakdown. All other things being equal, proteins are broken down at random but there is some evidence that larger molecules are more susceptible than small and that molecules with abnormal configurations, perhaps brought about by synthetic errors or the experimental incorporation of amino acid analogues, are degraded more rapidly (Yushok, 1974; Beauchene *et al.*, 1967; Bozouk, 1976). This molecular selection may involve subtle identification mechanisms, but more plausibly may just occur because abnormal proteins are intrinsically more susceptible to proteolytic attack. Certainly, cooked steak is more digestible than raw and *in vitro* it can be shown that molecular denaturation increases the susceptibility of molecules to proteolytic attack.

4.3 Cells in flux

Attention has already been directed to the dynamic state of many of the cell populations within the organism. There are, as already noted (Section 3.1), three main types of cell population; those that cease to proliferate at an early stage; those that only proliferate under unusual circumstances and those which turn over throughout the life of the organism. Even within the adult, non-growing organism, therefore, some cells will be turning over at a measurable rate and it is estimated that during each 24 hours of adult life each human being makes and sheds not less than one hundred, thousand million cells!

The cell systems which do turn over all the time are those which, due to their position or function, are likely to be subjected to whole-cell damage or destruction. Those cell systems which categorically do not turn over are ones concerned with a very specialized function, like muscle fibres, or those concerned with the processing and storage of information, like nerve cells. Even these systems, however, are not fully closed, for their intracellular contents turn over at a measurable rate. By observing the flow of freshly synthesized cytoplasm down the axons of nerve fibres for example, Weiss and Hiscoe (1948) calculated that 1/700th of the protein disappears every hour. This means that the whole intracellular cytoplasm of these cells turns over completely in approximately one month.

Turnover inevitably means that tissues not only retain a proliferative capacity but also that cells within them die and disappear. In some tissues, for example the skin and its appendages, the proliferative zone is separated from the zone of mortality and here cell death is usually preceded by the transformation of the cellular structure into a keratinized coat. Dead tissues are ultimately sloughed off from the skin surface. In other tissues there is a thorough intermixing of proliferative and dying cells and a major puzzle here is how cells are singled out for destruction. Correlated with cell death is an increase in the activity of acid phosphatase enzymes within cells and the breakdown products of the catabolic reactions catalyzed by these enzymes are usually removed by phagocytic cells, themselves rich in acid phosphatases. But it is not absolutely clear if hydrolytic enzymes and phagocytic cells are the causative agents of cell death or whether they are simply called into play to remove already dead or dying cells. It has been noted, for example, that there may be a lag between cell death and the increase in phosphatase activity in some experimental systems and alternatively, that the activity of the phosphatase enzymes may sometimes increase in the absence of cell death (Saunders, 1966). Except in zoned tissues, it is also very much an open question as to how selective the process is; for example, are malfunctioning or diseased cells removed preferentially?

Human red cells have been studied intensively as a model system of cell death despite the view held by some that these cells are atypical and, being without a nucleus, are essentially dead at the time of release into the circulation. Radiotracer studies have established that red blood cells live an average of 120 days. At the end of this time the cell is recognized in some way, removed from circulation and degraded. As they age, red cells change in specific gravity, become more sensitive to lysis and become more easily disrupted by mechanical treatment (Saunders, 1966). Some of the most interesting results have come from *in vitro* studies. The deterioration of cells in storage, for example, is correlated with a gradual decrease in ATP (Huennekens, 1960) and an increase in Pi. Deterioration can be retarded, however, by the addition of ATP to the medium. Furthermore, the prime action of ATP in retarding the ageing phenomenon seems to be as a source of readily metabolized energy. Bacterial cells also fail when the energy charge falls below 0.5. Hence it is tempting to suggest that cells may die through a running down of their metabolic processes and it is easy to see how such a process might in turn be influenced by genetic instruction or external disturbance. Against this interpretation, however, is the fact that the metabolic activity of mitochondria of nucleated cells may persist even after partial hydrolysis. For example, Houck (1973) has observed that fibroblasts subjected to lethal doses of irradiation quickly lose their ability to incorporate tritiated thymidine and uridine. After a lag of several hours these cells start to discharge lysosomal enzymes, release cytoplasmic enzymes into the extracellular space and become

Life Cycles

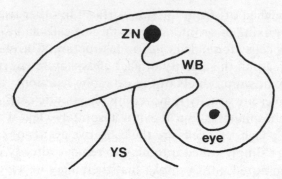

Fig. 4.1 Cell death in the normal development of the chick. WB = wing bud; YS = yolk sac; ZN = zone of necrosis.

permeable to vital dye. Even as the morphology of the cell becomes disrupted, however, one can still detect significant amounts of phosphorylation, oxygen consumption and glucose utilization. The inference here, then, is that cell death follows from an initial complete inhibition of transcription and translation rather than an inhibition of 'read-off' from parts of the genome concerned specifically with energy metabolism. To what extent these events simulate the normal process or to what extent observations on red blood cells reflect the genetically closed conditions of this enucleate system are matters for debate. It is very likely, of course, that cells may die in more than one way and that what are observed in the irradiated fibroblasts or the red blood cells *in vitro* are the extremes of a spectrum of possibilities. Indeed, cell death through accidental failure of any one of several subcellular systems, may be the rule rather than the exception. There are, however, situations, particularly during the early stages of development, when programmed cell death seems to occur and this is likely to depend on specific mechanisms of 'suicide' or 'assassination' (Saunders, 1966).

Glücksman (1951) identified three major roles for cell death in animal embryogenesis. First, there is *phylogenetic cell death* which is involved with removing organs and tissues only useful during larval life; for example, the pronephros and mesonephros of higher vertebrates, the anuran tail and gills and the larval organs of holometabolous insects. Second, *histogenetic cell death* plays a role in differentiation; for example the cell death involved in the remoulding of cartilage and bone. Finally, *morphogenetic cell death* is involved in folding and joining tissues; for example the necrotic zones that occur in the posterior part of the wing buds of chick and between the digits of developing mammal limbs

(Fig. 4.1). In that all these processes are necessary for normal development they must be programmed and carefully controlled. Extrinsic factors of systemic origin (e.g. thyroid hormone in anuran metamorphosis) or more local agents (e.g. unknown factors from sites in close proximity with the necrotic zones of the limbs and wings) seem to be required for the initiation of death. However, it is possible that these work through differential action on the expression of the genome, on the energy metabolism of the cells that are involved or, as in the fibroblast system, by completely switching off transcription and translation. Following initiation of the process, lysosomes, acid phosphatase enzymes and phagocytic cells become involved as in the adult tissues.

In summary, it is important to re-emphasize the intimate relationship between tissue turnover and cell death. This cell death may either occur by accident (in adult tissues) or by design (e.g. in embryonic systems and some adult tissues like skin). If tissue mass is balanced despite turnover there must be information flow between death rate and division rate, and perhaps vice versa. This we consider in more detail in the next section.

4.4 Control of cell turnover

The control of cell turnover is clearly important. However, the problem of analysis is difficult in that control might be effected in several, non-mutually exclusive ways. We know little, if anything, about the control of cell death and though there has been more investigation on the control of cell division, the state of the art remains confused, particularly for animal systems. Here, two questions have dominated the research programme: is control effected by specific controlling substances or by functional relationships; and is control by stimulation, inhibition or both? The one thing which has prevented any major breakthrough and solution of these issues has been a failure to isolate and characterize growth factors. This in itself, might argue against the involvement of specific, controlling factors but other, circumstantial evidence argues in their favour.

For several years, investigators have recognized the ability of liver to regenerate after partial hepatectomy, and the ability of one kidney to increase in functional mass after complete removal of the contralateral partner. These responses may be controlled by signalling substances or by the increased work-load inflicted on remaining tissue by the loss of other tissue (Goss, 1964). Despite some contradictory results, complicated cross-transfusion experiments favour the possibility that a circulating

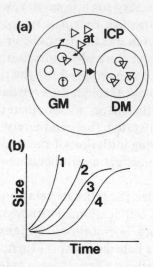

Fig. 4.2 A pictorial representation of the Weiss-Kavanau growth model.
(a) Tissue is thought to consist of generative mass (*GM*) which divides to
produce more *GM* and non-dividing, differentiated mass (*DM*). *GM* produces
a template molecule (*t*) which stimulates growth but which cannot escape
from the cells, and an anti-template (*at*) which inhibits *t* and which can
escape into the intercellular space (*ICP*). The rate of production of *GM* will
depend on the amount of *GM*, so $dGM/dt = k_1 \, GM$, where k = constant,
which predicts exponential growth — curve 1 in Fig.b. Weiss and Kavanau
suggest that the conversion of *GM* to *DM* will also depend on *GM*, so
$dGM/dt = k_1 \, GM - k_2 \, GM$, which yields curve 2 in Fig. b. The negative
feedback effect of the *at* is modelled by assuming that it is proportional
to C^x, where C = concentration of *at* and $x > 1$. Therefore, $dGM/dt =
k_1 \, GM - k_2 \, GM - k_3 \, C^x$, which yields curve 3 of Fig. b. Finally, catabolism
of *GM*, *DM* and *at* is taken into account as follows: $dGM/dt = k_1 \, GM -
k_2 \, GM - k_3 \, C^x - Ex$, where Ex = total excretion, which yields curve 4 of
Fig. b. Curves 3 and 4 are sigmoid and most real organisms grow in this
fashion.

factor may be involved in the control of proliferation (Bucher *et al.,*
1969). Furthermore, it has now been established that proliferation can be
stimulated in subcutaneous liver grafts which, because of their position,
do not function (Leong *et al.,* 1964; Lee, 1971). This is strong evidence
for the involvement of specific controlling factors. Of course, the same
method of control may not apply to every tissue and the matter is further
complicated by the fact that controlling substances could operate on the
basis of functional demands.

Proliferation is often caused to increase by the removal of tissue. This is true both in tissues in which cells divide only after wounding (see above) and in continuously dividing tissue (e.g. skin, Bullough and Laurence, 1960). If cell turnover increases after removal of 'something' then control under normal circumstances is likely to be by inhibition. There is also much evidence for stimulators because tissue extracts (e.g. of kidney) have been shown to stimulate proliferation (Weiss and Kavanau, 1957). Furthermore, Weiss and Kavanau (1957) have built an elaborate growth model in which control is by a combination of intracellular stimulators and extracellular inhibitors (Fig. 4.2) and the predictions from this conform closely to the growth response of animals under both normal and experimental conditions.

In plants the control mechanisms are more fully understood. Here, stimulation rather than inhibition seems to be the most important source of control (Steward and Krikorian, 1971). Specific, stimulatory hormones have been identified; for example, kinins which stimulate proliferation, and auxins and gibberellins which appear to stimulate both cell growth and division. There is also some experimental evidence for substances with anti-kinin, anti-gibberellin and anti-auxin effects. However, as with research in animals the major difficulty in validating this evidence has been in distinguishing between the non-specific, toxic effects of additives (which are of little interest) and the true inhibitory effects. Until inhibitors can be isolated and chemically characterized, and until their mode of action can be explained, their role in growth control must remain conjectural.

Within this apparent multiplicity of growth processes it would be convenient to find a common basis for control. It has already been suggested that proliferation may depend on nutrient or energy supply. It is not difficult to see how this could be controlled by either controlling substances or functional activity so that the 'energy basis of control' provides an attractive hypothesis (Calow, 1977b). Furthermore, as was originally proposed by Victor Twitty (1940), the regulated supply of nutrients to different tissues may form the basis of differential growth and thus of an important aspect of morphogenesis. However, control may be based on the regulated use of freely available metabolites, possibly through differential gene expression. The whole area is in a confused state and is in need of a critical experimental analysis.

5 On the adaptive significance of growing

Since fitness is measured in terms of the differential production of viable offspring (Section 1.5) it might be argued as a first approximation that selection should tend to favour those systems which convert the greatest proportion of their input nutrients into gametes. Some respiratory metabolism is required to power the process of replication and some structural and functional organization is needed to support reproduction, but why organisms should use energy to grow very big rather than to produce progeny is not immediately obvious. And yet some living things, like the whale and the sequoia do grow very large and during evolution the maximum size of organisms has apparently increased with the rise of each new phyletic organization (Fig. 5.1).

The apparent paradox between the actual evolutionary trend to get bigger and the theoretical need for competing systems to remain small and fecund, stems from the naive 'balance-sheet' argument which has been used above. We have ignored the fact that the accumulation of potential energy in parental protoplasm might, indirectly, facilitate the production of viable progeny. Whenever the returns, in terms of viable offspring, are greater from this indirect action than the returns from the direct production of gametes then the production of parental protoplasm will be favoured at the expense of reproduction. The fitness value of this 'catalytic effect' is difficult to quantify *a priori* and can most easily be judged by comparing the relative merits of different sized organisms under specific ecological conditions.

5.1 There are two ways to get bigger

In comparing the relative merits of bigness over smallness it is necessary to appreciate that there are at least two possible ways to become bigger

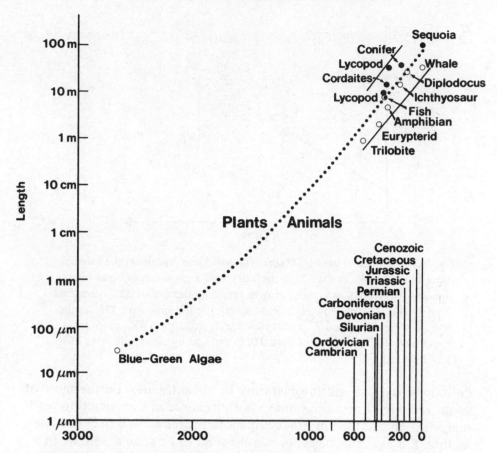

Fig. 5.1 Increase in the average size of organisms over the course of evolution. (With permission from Bonner, 1965, Fig. 19, p. 17. Reprinted by permission of Princeton University Press.)

(Calow, 1977a). A genetic change may either raise the size of an organism by increasing the growth rate or by increasing the length of the growing period (Fig. 5.2). In the first case the comparison is between individuals which at the same time reach different sizes whereas in the second case the comparison is between contemporaries which at the same time 'take different decisions' in terms of stopping growing. From the point of view of fitness both strategies must be judged separately.

Since the spread of a particular gene is like the accumulation of money in a bank account, by compound interest, and since early reproduction, brought about by an increased rate of growth to the reproductive condition, is analogous to the frequent collecting of interest, then the increased

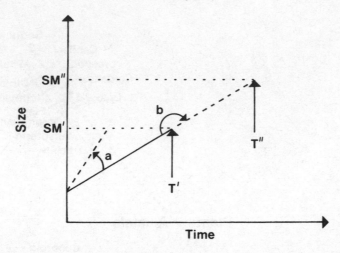

Fig. 5.2 Two ways to become bigger. The solid line represents the original condition with *SM'* as the size at maturity and *T'* as the age at maturity. Change *a* involves an increase in growth rate and means that *SM'* is reached more quickly. Change *b* involves lengthening the life-span with *SM"* as the new size at maturity and *T"* as the new age at maturity. (With permission from Calow 1977a, *Adv. Ecol. Res.* **10**. Copyright by Academic Press Inc. (London) Ltd.)

developmental rate will automatically be advantageous. Furthermore, it seems that bigger organisms have a better chance of surviving through to reproduction and then of producing more progeny when they become mature. For example, bigger organisms tend to be at an advantage in competitive situations involving direct combat and in terrestrial plants in the fight for light. Big animals are also better at warding off predators because they are usually stronger. Similarly, though it is unlikely that bigger animals can move any faster than smaller ones of the same organization, they will nevertheless move the same distance in fewer steps and therefore with less energy expenditure (Hill, 1950). Big animals therefore take longer than smaller animals to become exhausted and this will be advantageous in capturing food and in avoiding being eaten. Finally, since increases in volume cause a reduction in the surface-volume (S–V) ratio and since the body surface is an important region of material and energy transfer, size adjustments may have an important effect on the conservation of these resources. It must be remembered, of course, that though an increase in weight will cause a reduction in the S–V ratio and thus in the per gram loss of energy and matter from the body surface, it will invariably result in an increase in the absolute loss rate. S–V adjustments are therefore important in maintaining the constancy of the

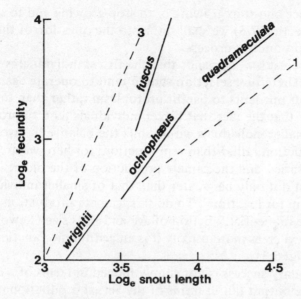

Fig. 5.3 Relationship between snout length and egg production in several species of desmognathid salamanders. The broken line marked 1 is the slope that would be expected if fecundity was proportional to body length (L^1). The broken line marked 3 is the slope that would be expected if fecundity was directly proprtional to body weight (L^3). Therefore, in these salamanders, the relationship between fecundity and body weight is no better than a direct proportionality, and this is typical of many animals (Calow, 1977a). The data in the figure is taken from Tilley (1968).

'milieu' on a per gram, not an absolute basis. Per gram levels are crucial in terms of tissue-water and heat and, significantly, the poikilotherms with poor cutaneous mechanisms of water conservation, become bigger in dry environments and homeotherms, whose proper functioning depends on the maintenance of a constant, above-ambient body temperature, get bigger in cold environments (Calow, 1977a). In general, therefore, bigger organisms have better chances of reaching a reproductive condition and then, as is demonstrated in Fig. 5.3, they tend to produce more gametes.

Similar arguments to those given above cannot be used as a justification of the other means of size increase; growing for longer. Here, extra growth usually occurs to the exclusion of reproduction. Therefore, the real decision is between stopping growing and starting to breed and vice versa. Sometimes this decision is absolute; sometimes it is modified by other factors. To begin with, though, we are interested in the most straightforward decision — to reproduce and stop growing or to stop growing and to start reproducing. Later (Section 7.5) we shall return to the question of the adaptability of the reproductive process.

The 'compound interest law' demands that the fittest individual is the one that breeds first. Therefore, selection should tend to operate against the conversion of input nutrients to parent protoplasm rather than to gametes. It is unlikely that the fact that bigger individuals tend to produce more gametes than smaller individuals would shift the balance in favour of protoplasmic production rather than reproduction; for here we are not comparing contemporaries, and the gamete production of the older, larger individuals must not only be greater than that of smaller individuals, but must also 'make up for lost time'. To do this, gamete production would need to be an exponential function of age and thus size (Lewontin, 1965; Calow, 1977a) whereas more usually it is an arithmetic function of these variables (e.g. Fig. 5.3).

Of course, evolutionary success is not simply judged in terms of potential reproductive output but in terms of the actual production of viable progeny. If, for example, as is plausible (see Section 7.3), reproduction puts the parent at risk and yet the parent has a better chance of surviving over a particular period of time than the progeny, then it is conceivable that the selective advantage might shift in favour of lengthening the life cycle. Indeed the parent need not grow through this period and may even shrink without affecting the wisdom of the 'decision' to delay breeding. Several species of British freshwater snail, for example, hatch in early spring and grow rapidly through summer to a size potentially large enough to support reproduction. Yet breeding is often delayed until the following spring presumably because large parents are better adapted to tolerate harsh winter conditions than small progeny (Russell-Hunter, 1961). This is the whole basis of seasonality — why living things often do not reproduce immediately they are able but at a time which best favours the transference of the genetic 'instructions' from parent to offspring. We shall discuss these question in a more rigorous way in the next part of the book.

5.2 Is there any benefit in adiposity and obesity?

From the last section it is easy to see that the production of parent protoplasm might have a selective advantage through its 'catalytic effect' on the production of reproducing progeny. On the other hand, building up the inert storage materials within the body makes no apparent

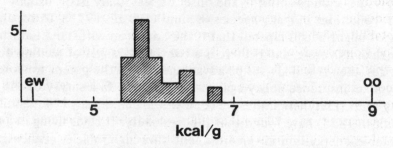

Fig. 5.4 Frequency distribution of energy equivalents (in kcal/g, multiply by
4.2 to obtain kJ/g) per ash-free dry weight for a variety of animals. (With
permission from Slobodkin and Richman, 1961). ew = calorific value of
egg white; oo = calorific value of olive oil. A more complete survey is now
to be found in Cummins and Wuychek (1971). Schroeder (1977) questions
the skewness of the distribution shown in the figure and suggests that it
might be represented as well by a normal distribution. However, the main
point with regard to the argument in the text is that the mean or median
of the observed distribution occurs towards the lower end of the expected
distribution of biological materials; i.e. from approx. 4.0 kcal/g (ca. 16 kJ/g)
for carbohydrates to approx. 9.0 kcal/g (ca. 38 kJ/g) for fats.

contribution to progeny production. As Slobodkin observed, 'there is a
selective advantage in increasing fecundity but not adiposity' (Slobodkin,
1962, p.7). In a resource-limiting environment excess energy could either
be converted directly to offspring or to materials which will ultimately
promote, not hinder, their production. Significantly, Slobodkin and
Richman (1961) found that the potential energy per gram of animal
tissue of 17 species from six phyla was skewed towards the lower limit
of the possible range of biological materials (Fig. 5.4). This implies that
fat, with a high energy value (37.8 kJ, 9.0 kcal/g), is a minor component
of animal biomass. Plant tissues have even lower energy values tending to
17.5 kJ/g in macrophytes and 19.0 kJ/g in algae.
    Like all generalizations, that of Slobodkin is in need of some elaboration
and correction. First, adipose tissue may have a direct functional effect
as in the insulation of homeotherms or in the buoyancy of plankters
(Wissing et al. 1973). Second, storage itself has an adaptive value as an
investment against periods of resource limitation that are interposed
between birth and maturity. For example, energy values are high in seeds
(Levin, 1974), rise prior to pupation in insects, and also rise prior to
migration in birds (Odum et al. 1965), whales (Brodie, 1975) and
termites (Wiegert and Coleman, 1979). Also, it is well known that

energy values increase prior to the onset of seasonally poor, usually
winter, conditions in many species (Schindler *et al.,* 1971). Third, and
finally, it might be anticipated that if the adult can withstand hardship
better than its young, and if there is a continual danger of feeding disturbances, then it will always be advantageous for the parent to store some
energy as an insurance policy against 'hard times'. In a survey of the equivalents of the Platyhelminthes we (Calow and Jennings, 1974; Jennings
and Calow, 1975) have found that the freshwater triclads living in an unpredictable, energy-limiting environment have high joule equivalents.
Alternatively, parasitic flatworms (flukes and cestodes) which live in a stable,
energy-rich environment have low joule equivalents. Ectosymbiotic species,
like the temnocephalids which live on the gills of fish, have a trophic ecology
intermediate between free-living species and obligate entoparasites, and
have intermediate joule equivalents. Exactly the same principles apply
to microbes living in environments of differing trophic stability in that
energy is usually stored when trophic conditions are unpredictable, and
not as was originally thought, only when there is an energy surplus
(Parnas and Cohen, 1976; Calow and Jennings, 1977).

Obesity in Western Man is clearly non-adaptive. It seems to have a
genetic and non-genetic component, both leading to the excessive
accumulation of adipose tissue. It also seems to depend on physiological
and psychological mis-adaptations. Physiologically, sufferers are unable
to match energy output and input and psychologically there seems to be
an inability to recognize satiety (Garrow, 1974). Chronic obesity is
clearly pathological and yet the metabolic apparatus which causes the
problem and which leads many people continually to watch their weight
may have once been part of a perfectly adapted physiological repertoire.
Our anthropoid ancestors may well have evolved in an unpredictable
nutritive environment where it would have been quite sensible to adopt
metabolic mechanisms of energy storage, like the triclads, as an insurance
against times of famine. Even in subsistence agricultural societies of today,
the heavy manual work of cultivating next year's crops frequently coincides
with reduced food availability. In these conditions, the ability to store
excess energy efficiently and to release it when needed for physical work
has survival value. In this context Dugdale and Payne (1977) have defined
the following ratio:

$$P = \frac{\text{energy stored as (or mobilized from) protein}}{\text{total energy stored or mobilized}}$$

For Western Man the modal value of P is between 0.05 and 0.10. Fat men
have a P value set at around 0.03 and lean men at 0.30. Using data for

Table 8 Seasonal patterns of weight change for subsistence farmers (data from Gambian farmers) derived from a computer model. (Results from Dugdale and Payne, 1977).

			Total wt. change (kg)	Δ Fat wt. (kg)	Δ Lean wt. (kg)
Fat men					
Cultivation period	start		59.0 ⎫	−3.4	−0.6
5 months	end		55.0 ⎭		
Harvest period			⎫	+4.8	+0.9
4 months			60.7 ⎭		
Rest period			⎫	−0.8	−0.1
3 months			59.8 ⎭		
Lean man					
Cultivation period	start		59.0 ⎫	−0.9	−2.2
	end		55.9 ⎭		
Harvest period			60.7	+1.5	+3.3
Rest period			59.4	−0.5	−0.8

food intake and activity schedules of Gambian farmers, Dugdale and Payne were able to compute total weight changes using various values of P. The results for 'fat' and 'lean' men are summarized in Table 8. 'Fat men' used mainly fat to balance their energy budgets but the 'lean men' drew upon their lean tissue, mainly muscle, to meet energy deficits. In consequence, the 'lean men' depleted their muscles at a time when these were needed for heavy work. In this substistence community, therefore, there is considerable selection pressure for low values of P and hence adiposity. In Western society an internal storage system has become redundant and unfashionable and yet there can only have been a weak selection against it. Consequently, we are left with a trait like the appendix, which was once of value, but which is now useless and which occasionally causes health problems.

Part three **Reproduction**

No organism is immortal. If it does not die 'naturally' it inevitably falls foul of accident, disease or predation. What has grown up ultimately dies out so for the sake of continuity there must be a process of reproduction — a link from one organismic cycle to another.

The essence of reproduction is the passing of the instructions for organization of the life of the race to a new unit. By reproduction the characteristic activity that is the life of the species is started again. (Young, 1971; p. 172).

Almost by definition, reproduction usually involves the release of a part from the parent individual. This part may be a multicellular fragment or more usually a single cell. In either case the part grows, principally by mitosis in metazoa, back to the parent state. The cycle is completed when the product of reproduction matures and reproduces again.

A qualitative description of reproduction will be given in Chapter 6 and several quantitative questions will be considered in Chapter 7. Each part of life's cycle has been subject to selection but since fitness expresses itself in the number of progeny produced by a parent it can be anticipated that reproduction and reproductive strategies will bear, in a particularly obvious way, the stamp of selection.

6 How organisms reproduce

6.1 Mitosis and meiosis

Reproduction generally involves two processes; copying and separation. Sometimes the visual copy is produced after separation but the pattern on which it is based must have been copied before. Ultimately this will involve copying duplicate strands of the genetic information in the nucleus of the cell. The most straightforward method of reproduction is therefore by mitosis (Section 3.1) and this is the method actually used in the binary fission of unicellular organisms (Fig. 6.1a). Mitosis is also the basis of asexual and vegetative reproduction in higher organisms, either being involved in the reproduction of the whole organism from separated fragments (Fig. 6.1b), or in the formation of miniature adults prior to separation and in their expansion afterwards (Fig. 6.1c). Sometimes single, germinal cells within an organism may, by mitosis, form new organisms, as in the formation of the redia and daughter sporocysts of parasitic termatodes (Fig. 6.1d), and in the formation of spores in some of the lower plants and fungi (Fig. 6.1e). Parthenogenesis, reproduction from unfertilized ova, (e.g. in dandelions and aphids), is also usually based on the mitotic proliferation of diploid germ cells of the parent which go on to develop by mitosis after formation.

There is another, more complex form of reproduction in which the new organism is the product of two parents. This is sexual reproduction. Usually, though not always, it occurs by the fusion of morphologically distinct cells, the gametes, derived from morphologically distinct parents. If the chromosome complement of successive, sexually produced individuals is to remain constant despite cellular fusion then there must be a mechanism which halves the chromosome number of the gametes from a diploid to a haploid condition. This reduction is achieved by a special division process called meiosis.

Chromosome numbers are reduced in meiosis by two consecutive

66

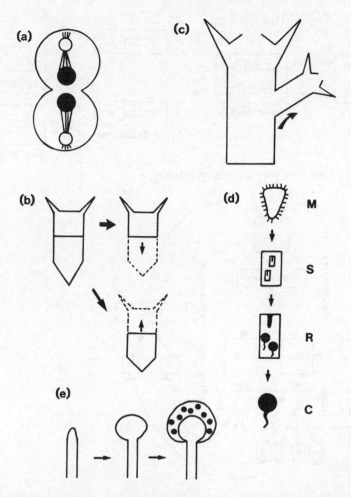

Fig. 6.1 Asexual reproduction by mitosis. (a) binary fission; (b) transverse
fission in triclads; (c) budding in *Hydra*; (d) asexual multiplication in the life
cycle of a trematode parasite (M = miracidium, S = sporocyst, R = redia,
C = cecaria); (e) spore formation.

divisions (Fig. 6.2), each analogous to mitosis but without nucleic acid
synthesis between them. For this reason it seems likely that the meiotic
mechanism has evolved by modification of the mitotic mechanism.
Unlike mitosis, however, the first division does not effect the separation
of the chromatids (i.e. the two daughter strands formed from the splitting
of each chromosome) but the separation of homologous chromosomes
(i.e. separate chromosomes which are thought in most cases to specify
identical characters). In meiosis the chromatids remain attached at the
centromere during the first division and only separate during the second

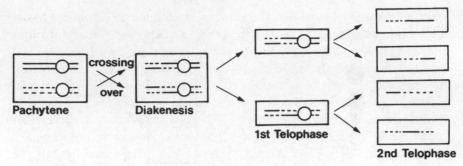

Fig. 6.2 Meiosis. See text for further explanation.

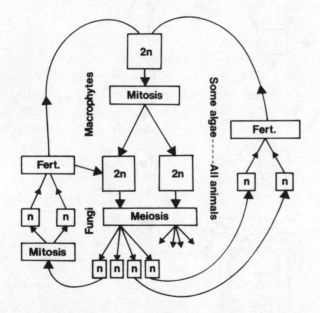

Fig. 6.3 Alternation of haploid (n) and diploid (2n) stages in the life cycle.

division. Often during the first reduction division an exchange of parts occurs between chromatids arising from homologous chromosomes (Fig. 6.2). This crossing-over brings about the mixing of genetic information and is, as we shall see later (Section 6.5), of profound evolutionary significance.

Meiosis thus produces haploid from diploid cells. Ultimately the haploid gametes fuse to produce a diploid individual and in consequence there is an alternation between diploid and haploid phases of the life cycle with mitosis interposed somewhere between. In fungi, mitosis is restricted to haploid cells, and in animals to diploid cells; but in most higher plants there is mitotic proliferation of both the haploid and diploid cells. Here haploid mitosis gives rise to the gametophyte (gamete-forming) tissue. This is illustrated in Fig. 6.3. The duration of the haploid and diploid phases may also vary according to the relative position of meiosis and fertilization in the cycle. In general, the diploid phase is dominant in animals and higher plants but the haploid phase is dominant in the fungi.

Diploidy has two evolutionary advantages over haploidy. First the diploid state, through heterozygosity and dominance, can 'shelter' and 'store' many hypofunctional genes that would otherwise have been eliminated in the haploid state. Hence, the potential variability of the gene pool is enhanced and the population is given limited ability to adapt, genetically, to changing conditions (Simpson, 1953). Second, by selection, diploidy slows down the spread of a recessive gene through a population. This is because a smaller proportion of the recessive genes is expressed in diploid populations, and this greatly reduces the force of selection operating on them. A lag of this kind can be of value in preventing a temporary switch in selection pressures from altering the population too hastily (Müller, 1949). These are definite long-term advantages but whether or not they provide reasons for the origin of diploidy and the suppression of the haploid state to an ephemeral gamete depends on whether they provide sufficient short-term benefits to be amenable to selection. On this basis, Simpson's hypothesis is more promising than that of Müller and perhaps the benefits of damping are simply a side-effect of the 'shelter' and 'storage' functions of diploidy.

6.2 Sexual dimorphism

Sexual dimorphism is a common but not a necessary feature of sexually reproducing organisms. Mating types in *Paramecium*, for example, can only be distinguished by the fact that members of one type can conjugate with those of the other. Furthermore, many plants and a few animals are hermaphrodite (Section 6.6). However, a division of labour between the two gametes, egg and sperm, and the producers of the two gametes, male and female, is the rule rather than the exception, particularly in the Animal Kingdom. The egg, with food reserves, is adapted to support the early stages of development and the sperm, with its flagellum, is adapted to reach the egg in order to initiate development. In primitive groups there is usually little difference between male and female but as the female becomes more adapted to receive the male and then to process the products of fertilization, and the male becomes more adapted to bring the sperm to the female, differences begin to occur and these may be reinforced by sexual selection.

As well as dimorphism between the sexes, there may also be a difference in their behaviour after copulation. In general, females become involved in post-natal care; sometimes the male may remain with the female as in Man; sometimes he may desert the female after copulation as in ducks, and sometimes he may be left to look after the reproducta, as in sticklebacks. Maynard Smith (1977a) has suggested a number of models to account for theses differences. These are based on discrete breeding but similar models can be built for continuous breeders:

(I) *Model in which success depends mainly on parental care*
Let V_1 = number of young which survive if only one parent cares for young and V_2 = number of young which survive if both parents care for young. If the male deserts a female he has a certain probability, p, of mating with another. If he does not desert the second mate, his fitness is $V_1 + pV_2$. Therefore male desertion is favoured if $V_1 + pV_2 > V_2$ or $p > (V_2 - V_1)/V_2$. Hence desertion is favoured when there is a good chance of finding a second mate and if a 'one-parent family' is almost as good as a 'two-parent family' in rearing viable offspring. The same argument might be applied to a female leaving a male to tend for the young. However, given a sex ratio of unity and the fact that there will be an inevitable delay between laying a batch of eggs and being ready to copulate successfully with another male, then p here is likely to be much lower than it was for the male. This therefore explains why female desertion is rarer than male desertion.

(II) *Model in which success depends on both parental care and the investment made by a female in laying*
P_0, P_1 and P_2 are respectively the probabilities of survival of eggs which are unguarded (D), or guarded (G) by one parent, or guarded by two parents, and $P_2 \geqslant P_1 \geqslant P_0$. As before, a male who deserts has a chance, p, of mating again. A female who deserts lays W eggs, and one who guards, w eggs, and $W \geqslant w$ since guarding takes time and energy. There are four possibilities: (a) $D♀$ and $D♂$ (common in many invertebrates) require $WP_0 > wP_1$ or the female will guard, and $P_0(1+p) > P_1$ or the male will guard; (b) $D♀$, $G♂$ (as in sticklebacks) requires $WP_1 > wP_2$ or the female will guard, and $P_1 > P_0(1+p)$ or the male will desert; (c) $G♀$, $D♂$ (as in ducks) requires $WP_1 > WP_0$ or the female will desert, and $P_1(1+p) > P_2$ or the male will guard; (d) $G♀$, $G♂$ (common in birds) requires $WP_2 > WP_1$ or the female will desert, $P_2 > P_1(1+p)$ or the male will desert.

Clearly in species with 'one-parent families' it will tend to be the male which cares for the young if the female has invested so much in egg production she cannot effectively do so, and this explains why female desertion is so common in fishes where fecundity is often explosive. Alternatively, it will tend to be the female that cares for the young if the deserting male has a better chance of remating than a deserting female. Surplus males will tend to favour female desertion and surplus females, male desertion. Wherever the young require continuous attention, as in many birds, it is likely that $V_2 \gg V_1$ and 'two-parent families' will be favoured. This is also true if 'two-parent families' guard better than one, as appears to be the case with geese and swans but not apparently with ducks (Maynard Smith, 1977a).

Sexual differences in morphology and behaviour can usually be referred back to differences in the gametes of the two sexes and often to differences between specific pairs of chromosomes – the so-called XY system. In *Drosophila,* Man and some dioecious plants, males are XY and females XX whereas in moths and birds the reverse is true. In the latter case the homogametic, male condition is usually referred to as WW and the heterogametic, female system as WZ. In fish and amphibia there are no visible differences between the chromosomes, but usually the males are genetically heterozygous and the females homozygous for sex determination. Genetic differences in these organisms must be manifest at a biochemical level rather than at the level of chromosome morphology.

6.3 Fertilization in animals

If gametes are produced in spatially separated parents, mechanisms must have evolved to allow them to come together for fertilization. In primitive, aquatic organisms either sperm or both sperm and eggs are released into the external environment. The spermatozoa are mobile and in this way make contact with the usually sessile egg. This kind of external fertilization is obviously very precarious and depends for its success on the release of vast quantities of gametes. Occasionally the process is facilitated by the release of sperm in close proximity to the eggs, as occurs in frogs and some fishes. A safer and more economical strategy, however, involves placement of the sperm inside the body of the female and internal fertilization. This necessitates bringing sexually mature adults together at the right time.

There is a tremendous variety in the events leading up to internal fertilization (for comparative treatments see Wendt, 1965 and Street, 1974) but two characteristic features are always present: attraction and copulation. Attraction occurs by one sex signalling its position to others using either chemical, auditory or visual means. Many insects, for example, release olfactory pheromones which have the twin advantages of being capable of dispersing over long distances and yet being inconspicuous to predators. Birds and some insects employ the auditory channel to produce a species-specific mating call. These travel long distances but probably not as far as pheromones. In addition, 'chirping' may be energetically more expensive than the release of a minute amount of chemical, and may alert potential predators. However, an auditory signal is probably much less subject to environmental vagaries, like wind, than an olfactory one. A beautiful example of visual stimulation is given by fireflies. These manufacture light flashes with the aid of photochemical equipment at the tip of the abdomen. Flashing appears to make these organisms very conspicuous and thus vulnerable — far more so than olfactory and auditory signals — but fireflies seem very distasteful and are rarely attacked.

Copulation involves sperm transfer, usually in a liquid semen, and fertilization is the act of fusion of gametes leading to the formation of the zygote. In natural populations there is no point in attraction and copulation without fertilization and indeed the whole process can be deemed a failure — a waste of time and energy — if progeny do not result from it. For this reason there must be careful control between the 'ripening' of the gametes within the female, her preparation for egg-laying or pregnancy and her responsiveness and/or attractiveness to the male. This requires considerable co-adaptation between the physiology, morphology and behaviour of the two sexes and one problem here is in trying to understand how this could have originated by chance, all at once. Indeed, it seems almost certain that a step-wise process must have been involved. Maynard Smith (1975) gives the following example. When oysters and sea urchins release gametes into the water they also release a chemical which stimulates conspecifics to do likewise. It is no good releasing a signalling substance if your neighbours will not respond. He suggests, however, that these signalling substances originally had other functions — for example, nourishment of gametes. Should the presence of these nutritive substances then have evoked gamete release from neighbouring individuals, reproductive benefits would have been conferred upon those organisms with the substance. Hence this co-adapted trait could have evolved by a process of function displacement, from nutrition to communication, and this could also provide a plausible explanation of the evolution of other co-adapted traits involved in the mating process.

There is no better example of this co-ordination in mating systems than the well-documented sexual cycle of mammals (Everett, 1961). We are apt to take the human cycle, with its 28-day frequency and few pregnancies, as a model of this process; but it is not very representative of the situation in natural, mammalian populations where the non-pregnant cycle is a rarity. Mammals, particularly small, short-lived ones, must safeguard against not becoming pregnant at the appropriate time in the cycle, or recover and re-cycle as soon as possible after failure. Some female mammals ovulate spontaneously at specific times. To increase the probability of fertilization after ovulation the female gives specific, often olfactory, signals that she is 'on heat', and only then is she receptive to male court-ship. This occurs, for example, in the bitch and the cow. However, as information becomes available on more and more natural populations, it appears that induced ovulation, where egg release occurs under the influence of copulation, is far more common (Conaway, 1971). Both spontaneous ovulation without copulation and induced ovulation without fertilization may lead to a period of pseudopregnancy, which involves the preparation of the uterine wall for an implantation that will never take place. During this time there is no possibility of a further successful copulation. Though difficult to quantify, it is probably fair to say that copulation without fertilization is very rare in nature, and this means that induced ovulation ensures that the female is always ready for fertilization and that little time is wasted in pseudopregnancies. If this is so, the question arises as to why spontaneous ovulation should ever have evolved in the first place. One possible explanation is that it helps to spread out and randomize ovulations of females in a population or groups of mammals (Conaway, 1971). This would be advantageous when offspring require an extended period of concentrated care.

6.4 Events leading to fertilization in plants

The problems posed by fertilization in plants are in many ways similar to those posed in the Animal Kingdom, but are more acute in that plants, particularly the macrophytes, do not move. As usual, the primitive mechanism involves the free release of one or both gametes into the aqueous environment. Even in some terrestrial plants, like the fern, the male gamete swims in a watery film, trapped between the gametophyte thallus and the soil, to the female gamete. Transport of this kind is not appropriate, however, for the fully terrestrial, vascular plants, where the main problem is in getting the male gamete, encapsulated in a protective pollen grain, through 'dry air' to the female pistil.

In grasses and cereals, pollen is literally thrown to the wind.

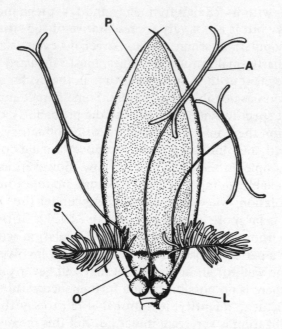

Fig. 6.4 Flower of Meadow Fescue with lemma removed to reveal feathery
stigma (redrawn from Rendle, 1930, by L.J. Calow). This adaptation is
typical of the grasses and sedges. A = anther; L = lodicule; O = ovary;
P = palea; S = stigma.

Consequently, the chances of fertilization are slim, and to compensate
vast amounts of pollen are produced. In addition, the ovaries are topped
with feathery plumes — the stigmas — to facilitate the collection of pollen,
and the flowers themselves are raised on stalks (Fig. 6.4). Often, however,
flowering plants adopt a less wasteful system involving animal vectors —
usually insects, but sometimes animals such as birds and bats — and here
the evolution of means of attracting the vectors becomes important.
Colour, shape and smell are well-known floral modes of attraction
(Faegri and van der Pijl, 1966), but the presence of nectar is also an
important criterion (Heinrich and Raven, 1972). For efficient out-
crossing, sufficient nectar must be produced by one plant for attraction
but not too much to discourage the visitor from going on to other plants.
The specific amount of nectar produced by a flower must be related to
the energy requirements of the pollinators. Small plants do not attract
large pollinators who would have to work too hard for their rewards, and
species producing large amounts of nectar are often adapted to exclude
small pollinators. For example, *Centopogen talamencis* in the mountains

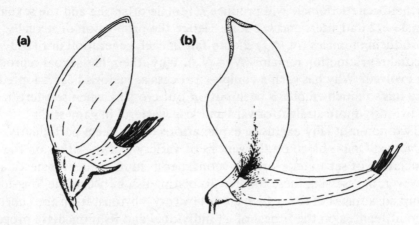

Fig. 6.5 Flowers of the *Erythina*. (a) *E. subumbrans*, pollinated by Sunbirds with a short beak. (b) *E. umbrosa*, pollinated by Hummingbirds with a long beak (redrawn from Doctors van Leeuwen, 1931, by L.J. Calow).

of Costa Rica is visited by the large humming bird, *Eugenes fulgens spectabilis*. This plant produces large amounts of nectar, and to exclude small pollinators, like bees, has a long, tubular corolla (Heinrich and Raven, 1972). Similar adaptations are to be found in the *Erythina* and these are illustrated in Fig. 6.5.

6.5 Why sex?

Sexual reproduction occurs even in bacteria and is probably very ancient. There are, in fact, no major groups entirely devoid of sex. On the other hand, as a means of passing on genetic information from one generation to another, it seems very precarious. Sexually mature organisms have to be brought together and ultimately the gametes have to fuse to form viable zygotes. In comparison with asexual reproduction, meiosis, attraction, copulation and fertilization take more time and energy, thus reducing the potential returns from the reproductive process. Furthermore, in terms of fitness sexual reproduction has roughly a 50% disadvantage with respect to asexual reproduction (Maynard Smith, 1971 a and b), for one sex cell only contributes toward 'half an individual' whereas each asexual spore forms a whole individual. Consider, for example, a population in which there are N_p parthenogenetic females, N_s sexually-reproducing females and N_s males; then the total population, N_t, is given by $N_p + 2N_s$. If all females have the same reproductive potential (k) then the parthenogenetic female will produce k female offspring and the sexual female $k/2$ daughters and $k/2$ sons. Hence the proportion of sexually reproducing females (N_s/N_p) should fall in each generation until only asexual reproduction remains $(N_t = N_p)$. Why, then, has sexual reproduction evolved? Why has such a complex process as meiosis been adopted? Why has so much emphasis been put on out-crossing when self-fertilization is obviously more straightforward and less wasteful in gametes?

Two non-mutually exclusive explanations of sex seem particularly attractive. One is based on the genesis of variety and the other on the association of separately formed, spontaneous mutations. It is necessary, however, in assessing these theories to distinguish between the long-term group advantages of sex, which are often very obvious, from the short-term influences on the fitness of an individual and its immediate progeny. The latter are not so immediately obvious, and in order to be advantageous they must, as a minimum, pay the cost of meiosis. On this, Treisman and Dawkins (1976) point out that the exact value of the meiotic cost depends crucially on the value of the sex ratio, r. Only when $r = 1$ is the cost of meiosis twice as much as that of parthenogenesis. In the general case the number of daughters (D) produced by a sexual female is $[1/(1+r)]k$. Hence as r becomes smaller, the closer D tends to k, and the smaller becomes the cost of meiosis. It might reasonably be argues that in the evolution of sexual reproduction from a parthenogenetic stock there would be more females than males, but this does not explain the continued persistence of sex in species where $r = 1$ (which is very common). In any event, if we can think of circumstances in which sex would be at an advantage when r is greatest, then we can be more confident in the plausibility of a conventional, neo-Darwinian explanation of the origin of sex. Hence we shall continue to search for a minimum, two-fold advantage of sexual over asexual reproduction.

In asexual reproduction the only method for introducing variability is through mutation, and yet mutation rates are usually quite low (Table 9). In sexual reproduction, however, the joint effect of cross-over (which generates new combinations of genes on the same chromosome) and independent assortment (which changes combinations of chromosomes) is to produce large genetic diversity even without mutation. Consider a genetic system consisting of just two loci with two alleles (A, a, B, b): then with complete intermixing, nine genotypes are possible: AABB, AABb, AAbb, AaBB, AaBb, Aabb, aaBB, aaBb, aabb. In general the total number of diploid genotypes from a species of n loci is 3^n. Since species like *Drosophila* and Man contain in the order of 10 000 loci

Table 9 Sample of spontaneous mutation rates (from Sager and Ryan, 1961).

Bacteria (*Escherichia coli*)	
lac⁻ →lac⁺	2×10^{-7} per cell division
Algae (*Chlamydomonas reinhardi*)	
Streptomycin sensitivity	
Str—s →Str—r	1×10^{-6} per cell division
Fungi (*Neurospora crassa*)	
inositol requirement	
inos⁻ →inos⁺	8×10^{-8} mutant frequency in asexual spores
Higher plant (*Zea mays*)	
Shrunken seeds	
Sh →sh	1×10^{-5}
Animal	
(*Drosophila melanogaster*)	
Yellow body	
Y →y ♂	1×10^{-4} mutant frequency
♀	1×10^{-5} per gametes
Mus musculus	
Piebald	
S →s	3×10^{-5}
Homo sapiens	
Normal →haemophilic	3×10^{-5}
Normal →albino	3×10^{-5}

bearing numerous alleles, the total number of conceivable genotypes is astronomical. Hence one major advantage of sexual over asexual reproduction is its ability to generate a greater degree of genotypic and thus phenotypic variability in a population. This will certainly be advantageous in the long-term but on what basis will it produce a short-term advantage sufficient to over-ride the cost of sexual reproduction? Williams and Mitton (1973) and Williams (1975) have described three model situations where the advantage might apply:

I '*Aphid-rotifer model*'. Here a diploid species propagates asexually through the growing season in small, separated habitats. At the end of the season propagules are produced sexually or asexually and disperse to new habitats. Consider one of these new habitats colonized by *m* descendents of an asexual parent, all with the same genotype, and *n* of a sexual parent, all with varying genotypes. *m* and *n* have the same mean, λ, over several habitats. The colonists all propagate asexually to fill the new habitat, and the fittest clone displaces the other. The outcome of this competition depends not on the original abundance of the genotype but on its ability to propagate successfully under the new conditions. Since the sexual form has n genotypes then it has n chances of giving rise to the winner whereas the asexual form has only a single chance. Williams and Mitton (1973) showed by simulation experiments that as the number of asexual generations in the new habitat and the value of λ increase, then the proportion of organisms carrying a sexual genotype rises. The cost of meiosis is more than met when this proportion is greater than 0.67.

Maynard-Smith (1971b) developed a very similar model to this, but arrived at a more stringent requirement for the success of the sexual line: 'All correlations between selectively relevant features of the environment [have to] change sign between one generation and the next'. For example, a wet environment would have to become dry and vice versa. The difference between the outcome of the two models hinges on the fact that the model of Maynard-Smith assumes that every colonist is produced randomly and independently from all those available, whereas the model of Williams and Mitton assumes specific, non-random colonization. Therefore, the Maynard Smith model applies when colonies are available for considerable, repeated invasion whereas the Williams and Mitton model applies to situations where invasion is rare.

II *'Elm-Oyster Model'*. Here it is proposed that such a vast number of progeny are produced per parent that many will colonize a site on which only one can survive. The subsequent internal competition will select a genotype which is optimally adapted for this condition, so abundance of the original propagules is of little account. Simulation experiments showed that when the number of original propagules was greater than 700, there was sufficient advantage in sexual reproduction more than to pay the maximum cost of meiosis.

III *'Strawberry-Coral Model'*. Here it is argued that by asexual reproduction the organisms can spread vegetatively like strawberries and corals to the limits of the area to which they are optimally adapted. If, however, they could also reproduce sexually, then seed could be distributed beyond these limits and could reach and colonize new habitats. A quantitative argument has not been worked out for this model.

The Williams-Mitton models, though plausible and instructive, have two disadvantages. First, they do not explain the origin and particularly the maintenance of sexual reproduction in species with low fecundity, like birds and mammals, and second, they deal with the progeny of, at most, two parents (Treisman, 1976). The first objection is telling, since even though birds and mammals might have 'inherited' the sexual state from a more fecund protochordate ancestor, we might have expected whole-sale return to the asexual habit, and yet this does not occur (Emlen, 1973). The second objection is also important, since as soon as the possibility of progeny from different parents is allowed into the arguments, then the potential variability in the asexual stock is raised, and the chances of sexual supremacy become correspondingly decreased.

Treisman (1976) has attempted to save the situation with a more general and plausible model. The essential features of this are: (1) that environmental variation is frequent and not exceptional and (2) that the adaptation to a major environmental variable is likely to be polygenic. Thus the adaptation to a variable like temperature might depend on a polygenic system of N loci, at each of which there are two homologous genes. The total effect of these $2N$ genes is represented by a score, G, consisting of the sum of the contributions of the component genes. For simplicity, Treisman assumes that:

(1) each allele makes an equal contribution to G;
(2) that at each locus there may be a 'high' allele which contributes +0.5 to G, and a 'low' allele which contributes −0.5;
(3) that the interaction between alleles is additive such that homozygotes may score +1 and −1 and heterozygotes 0 (therefore if $N = 3$, G ranges from −3 to 3, if $N = 4$, G ranges from −4 to 4 etc);
(4) that sexual reproduction allows independent segregation at each locus.
Ignoring the males, it is further assumed that there is a set of G's within which the G of a particular female must be if she is to be fertile.

Consider two situations:

(A) *Habitat in which a variable, e.g. temperature, is stable over a long period of time (optimum G constant = G^1)*
In this habitat there are initially equal numbers of sexual and partheno-genetic females. The parthenogenetic females all have the same genotype with a score (G_p) within the range of G^1. The parthenogenetic females will therefore breed, and since they produce twice as many offspring as the sexual forms they will be victorious. Furthermore, their advantage will be even greater because whereas all the young will have G s in the region of G^1, this will be true of only a portion of the daughters of the sexual females.

Life Cycles

Table 10 Number of parthenogenetic (n_p) and sexual (n_s) females over successive generations suffering a long-term fluctuation in environmental temperature. Parthenogenetic females have a genotypical score (G_p) of zero and asexual females begin with a mean score $(\overline{G})$ of zero. The high and low alleles are initially equally frequent at each locus in the asexual organism and are binomially distributed. The original population is adapted to the starting temperature and females in the range -1 to $+1$ are fertile. See text for further explanation. (Data from Treisman, 1976).

Generation	n_s	n_p	n_s	n_p	$\overline{G}$	Range of G's in sexual females
0	16	16	16	16	0	-1 to 1
1	28	64	56	128	0	0 to 2
2	39	256	156	1024	0.26	1 to 3
3	31.1	0	249	0	0.73	2 to 4
4	11.4	0	183	0	1.37	2 to 4
5	10.4	0	332	0	1.68	1 to 3
6	20.3	0	1 298	0	1.53	0 to 2
7	40.5	0	5 187	0	1.53	-1 to 1
8	33.4	0	8 541	0	1.19	-1 to 1
9	43.7	0	22 350	0	0.98	$-$

(B) *Variable habitat, e.g. temperature varies from one generation to another (G^1 varies)*

In this situation G^1 varies from one generation to another. Since G^1 is likely to be a narrow range it is probable that it will not always include the G range of the parthenogenetic stock. Once out of the range the parthenogenetic form will not breed. Alternatively, each G^1 is likely to include the G values of sexual forms because of the increased potential for variability. Hence, provided the shift away from G_p is frequent enough or long enough then the sexual form can have the advantage. A simulation model is illustrated in Table 10.

It is possible, therefore, on the basis of a series of models differing in degrees of plausibility and generality, to explain the origin and persistence of sex on the basis of the genesis of variety. Let us now consider the possible involvement of the other potential advantage of sexual reproduction; its ability through crossing-over, assortment and gametic fusion to bring together advantageous mixtures of genes and, in particular, to bring together advantageous mutations that developed independently in different genotypes. In asexual forms, two advantageous mutations can only come together in one individual if they arise in the same ancestral

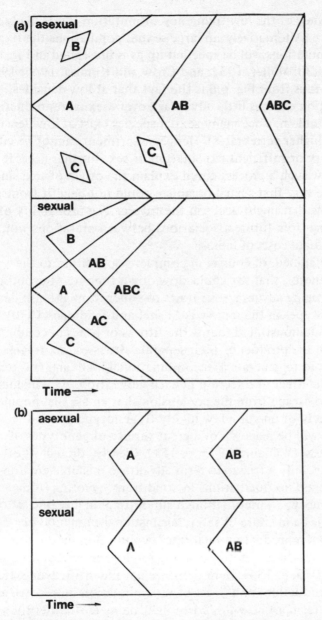

Fig. 6.6 The coming together of advantageous mutations (A, B and C) in a sexual and an asexual population, (a) for a large population and (b) for a small population. The smaller the population the lower the chance of advantageous mutations occuring simultaneously in the population and the greater the chances of the loss of any one mutation. (After Crow and Kimura, 1970).

genotype, and in view of the low frequency of mutation this is likely to be a rare occurence. Alternatively, in large sexual populations the fixation of advantageous mutations will be speeded up as is illustrated in Fig. 6.6 (see also Fisher, 1930; Müller, 1932; and Crow and Kimura, 1965, 1971). However, also obvious from Fig. 6.6 is the fact that at low densities, sexual recombination etc. has little advantage over asexual reproduction and this poses a problem, since many sexual species exist at low densities (e.g. many of the higher vertebrates). Hence, assortment cannot be cited as either a necessary or sufficient explanation of sex. Furthermore, it is difficult to see how such a process could explain the origin of sex, since the progeny of the very first sexual organism could not benefit from genetic associations that might occur in the future. The probability of advantages accruing from future associations between mutations would not pay the immediate costs of meiosis.

It might also be argued, of course, in complete opposition to the 'association hypothesis', that sex could slow down the rate of evolution since as well as bringing advantageous traits together, it might also destroy advantageous genotypes in the same way. Eshel and Feldman (1970) have convincingly demonstrated that if the fitness of genes in combination is ever greater than the product of their separate fitnesses, sex retards evolution. Sex is said to generate a 'recombinational load' and this tends to work against the effect of selection in each generation. As Williams writes, 'There is no escape from the conclusion that an asexual population should consist largely of one or a few highly fit genotypes while the sexual population will be made up of a great variety of genotypes of lower average fitness' (Williams, 1975, p. 151). Clearly, though, the asexual condition is only a true long-term advantage in stable environments where there are no fluctuations in conditions. Quoting from Williams again, 'genotype variety [though suboptimal in any one set of conditions] provides a measure of safety against environmental uncertainty ... . Sexual reproduction facilitates evolution indirectly by making extinction less likely, not by making phylogenetic change more rapid' (Williams, 1975, p. 154). Here then, we have identified an extra cost of sexual reproduction, or at least of recombination in meiosis, and a way, in the long-term, of how this cost might be met. Whether the equation balances, however, will depend on the frequency and intensity of environmental change, and there are those who consider that changes sufficient to shift the balance in favour of the evolution of sex are likely to be rare (see Aphid-rotifer model and Treisman's model above).

Returning, finally, to the possible advantages that might arise from sex in terms of the association of genes, there is the possibility of genetic 'hitch-hiking' (Strobeck *et al.* 1976; Maynard-Smith 1977b). Here, genes bringing about recombination are favoured since their presence increases the production of selectively advantageous types of gamete with which they remain associated. Suppose that there are three linked loci, a,b and c, in a population with a balanced polymorphism for alleles a and A (e.g. Aa is fitter than AA and aa). A favourable mutation (B) occurs at b. This will increase in frequency for a time but if there is no recombination the spread will be interfered with by selection and B will never go to fixation. Now suppose alleles C and c control recombination such that recombination between A and B occurs only with the homozygous CC. Sooner or later aBC will arise and will increase in frequency. In this case C, the gene for recombination, hitches a genetic lift on aB and goes to fixation, provided, that is, it is closely enough linked with these loci — i.e. it can hang on tightly enough. How tight the association needs to be is discussed in Strobeck *et al.* (1976); but in any event there is definite potential here for indirect selection for recombination even in the face of the disadvantage discussed above.

In summary, then, the major evolutionary significance of sex, both long- and short-term, would seem to be in the genesis of variety, and the association of mutations is probably a less important secondary advantage. The 'hitch-hiking hypothesis' has a lot of appeal but will require inspired genetic experimentation for ultimate validation or refutation. Indeed the resolution of many of the issues discussed above offers a serious experimental challenge for the future. Whatever the outcomes of these analyses, however, one conclusion is beyond doubt. This is that sex has had a profound effect on the genetics of populations. Because of sex Fig. 1.1, for example, becomes only a very simplified version of genetic transmission. Genetic lines of transmission are not isolated but thoroughly interdigitated. The things that are transmitted are not intact genetic programmes, but individual genes which are brought together within single organisms from a gene pool. The extent to which genes are transmitted to the future pool is a true measure of their fitness.

6.6 Why two sexes?

Sexual reproduction involves meiosis, separation of the haploid gametes from the parents and subsequent fusion to form a diploid zygote. In Section 6.2 we saw how the gametes are often produced by separate parents, but this is not a prerequisite of sexual reproduction and indeed a few animals and many species of plant are hermaphrodite. Here, one parent is capable of producing gametes which, in principle, can fuse with each other to form a viable zygote.

The advantages of hermaphroditism are numerous. For example, when the density of parents is low, or when they are sluggish or sessile (as in plants) there is a higher probability of cross-fertilization in hermaphrodites because with hermaphroditism there are twice as many chances of a successful mating contact (Altenburg, 1934 and Tomlinson, 1966) than with gonochorism (two separate sexes). Furthermore, hermaphroditism means that effectively there is no cost of meiosis as discussed in the previous section (Maynard-Smith, 1971 a and b). The question therefore arises as to why hermaphroditism is not more common, particularly in the Animal Kingdom. The only solution can be that there is a cost associated with hermaphroditism and Heath (1977) has ascribed this to the energetic cost of producing and maintaining two reproductive systems within a single individual. If the total energetic cost of reproduction per parent is R and the cost of producing and maintaining the male and female system is respectively m and f, then the balance left for progeny production in a pair of hermaphrodites and a pair of opposite sexed gonochorists is:

$$[R - (m+f)] + [R - (m+f)] = 2R - 2(m+f)$$

(a) for the hermaphrodites $[R+(m-f)] + [R+(m-f)] = 2R + 2(m-f)$

(b) for the gonochorists $\quad [R-m] + [R-f] = 2R - (m+f)$

Hence, assuming simple summation, the gonochorists will always have to spend more energy on gamete production than the hermaphrodites. In situations where the probability of successful mating contacts between adults is high, then gonochorism should be at an advantage. Heath reasons that since mating success is likely to be dependent on population density in mobile species, then at low density hermaphroditism will be favoured, whereas at higher density gonochorism will be favoured (Fig. 6.7). Similarly, as the mobility of species increases, the cost of hermaphroditism is likely to become more important than the cost of separate sexes. This would explain the widespread occurrence of hermaphoditism in higher plants and in slow-moving animal phyla like the Coelenterata, Platyhelminthes and Mollusca. In the latter phylum it is interesting to note that gonochorism comes to predominate only in the more active Cephalopoda.

The cost of hermaphroditism can be reduced, of course, by allowing the male and female systems to adopt common glands and ducts. Here, the cost of producing and maintaining the male and female systems in

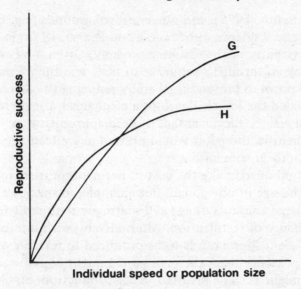

Fig. 6.7 The relative fitness of hermaphroditism (H) and gonochorism (G) as population density increases and as the speed of the individuals in the population increases. (After Heath, 1977).

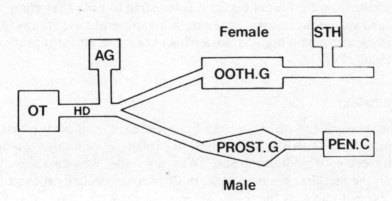

Fig. 6.8 Hermaphrodite system in a gastropod snail. AG = albumen gland; HD = hermaphrodite duct; OT = ovotestis; OOTH. G = oothecal galnd; PROST. G = prostate gland; PEN. C = penial complex; STH = spermatheca. (With permission from Calow, 1978a).

the hermaphrodite is not twice that in two separate gonochorists. In the Gastropoda, for example, a common gonad, the hermaphrodite gland or ovo-testis, may serve both the male and female functions (Fig. 6.8). The danger here, of course, lies in the possibility of self-fertilization which on theoretical (Maynard-Smith, 1975), and observational grounds (e.g. Barnes and Crisp, 1956; Gee and Williams, 1965; Giese and Pearse, 1975) is known to reduce the viability of the resulting progeny. Often, however, the hermaphrodite may go through a sequence of male and then female activity, a condition known as protandry, thereby reducing the dangers of self-fertilization. Provided the female develops a mechanism for sperm storage this would not reduce the advantage of hermaphroditism in reducing the cost of meiosis, though it would prevent any advantages that might derive from reciprocal copulation.

An alternative method of reducing the cost of hermaphroditism might be through reducing the size of one gonad. For example, in species with external fertilization large amounts of egg and sperm are required to ensure a reasonable chance of fertilization. Alternatively, with internal fertilization only sufficient sperm needs to be produced to fertilize a limited number of eggs. Hence, here the male gonad might be reduced and the extra energy might then be diverted to egg production. Similarly, when the capacity of the female system is limited, for example with viviparity where only a small number of progeny may be dealt with at once, the output of the ovary may be reduced and the balance shifted to sperm production. In this context it is interesting to note that many marine and viviparous invertebrates are hermaphrodite — perhaps to capitalize on the possibility of transferring excess energy from ovary to testis (Heath, 1977).

6.7 Conslusions

It has been possible to offer a series of models, each built on the basis of conventional neo-Darwinian principles, to explain reproductive features of the life cycle; 'Why dimorphism?' 'Why sex?' and 'Why two sexes?' However, the fact that the models work does not necessarily mean that it is by these methods that the reproductive features have come into existence and evolved. In any complex situation there is likely to be more than one equally satisfactory solution to the same problem. What is needed in this aspect of life cycle studies is experimental probity. It should be possible, for example, to determine under contrived conditions the relative merits of one-and two-parent guarding, and it ought to be possible to measure the cost of producing and maintaining the reproductive apparatus in gonochorists and hermaphrodites. On the question of 'Why sex?' what seems to be needed, particularly urgently, is more information on just how variable conditions are in nature and on just how variable they need to be to promote sex.

7 Quantitative aspects of reproduction

7.1 When to reproduce and by how much?

In Section 1.5 we defined fitness in terms of replicative capacity. A more subtle definition of fitness is in terms of the extent to which a particular trait comes to monopolize the resources available to it in a given habitat (Lotka, 1922). Only when it can be assumed that these resources are essentially unlimited does the initial definition of fitness hold.

In the simple case where resources are assumed to be unlimited, fitness can be expressed as the capacity of the bearers of a particular, genetically determined trait to increase in numbers. Under such conditions the change in size (N) of the population can be expressed by:

$$dN/dt = rN \qquad \qquad 7.1$$

where: r = the innate capacity for increase. Integrating equation 7.1 gives:

$$N_t = N_o e^{rt} \qquad \qquad 7.2$$

where: N_t = number of organisms at time t; N_o = initial numbers; e = base of natural logarithms. How is r related to the coefficient of selection (C) and the selective value ($W = 1-C$) defined in Section 1.5? Consider two sub populations (i and j) bearing genetically determined traits which differ in fitness. The bearers of each trait have different capacities for increasing such that:

$$N_{it} = N_{io} e^{r_i t}$$

$$N_{jt} = N_{jo} e^{r_j t}$$

Then:

$$N_{it}/N_{io} = e^r it = b_1$$

$$N_{jt}/N_{jo} = e^r jt = b_2$$

Now if both subpopulations begin at the same size ($N_{io} = N_{jo}$), then from Section 1.5 the selective value may be defined as follows:

$$W = b_1/b_2 = e^{r_i - r_j}$$

Hence the relationship between r and W can be summarized by:

$$\log_e W = r_i - r_j$$

In what follows we assume that we are considering sub-populations which carry traits differing from the stock population such that r_i (for the sub-population) differs from r_j (for the stock). Hence r_j is a constant and the value of W varies only with that of r_i. Here, then, it is sufficient to use r_i as a measure of fitness. It should be noted, however, that this approach does introduce some obscurity since it shifts the emphasis from genes to individuals (Stearns, 1977). In so doing it plays down the importance of sex and gene mixing, the concept of the gene pool and the definition of fitness in terms of gene frequency (Section 1.5). We are trading off some realism, therefore, for a way of judging the fitness of measurable population traits and this must be kept at the back of our minds in the analysis that follows.

Another problem is that equations 7.1 and 7.2 apply only to continuously breeding populations with a stable age distribution and these conditions are rare. More usually, breeding is seasonal and N increases in a stepwise fashion, not continuously. Under these conditions the following equation (Caughley 1967) applies:

$$r^1 = \frac{\log_e R_o}{T} \qquad\qquad 7.3$$

where: r^1 is roughly equivalent to innate capacity for increase; R_o = number of progeny per head of parents which survive to reproduce (i.e. N_t/N_o or $P \times B$; B being mean natality and P being mean survival of progeny to reproduction); T = time from one breeding period to the next. Clearly, r^1 is sensitive to R_o and to generation time, T. However, because of the logarithmic expression of R_o, small changes in T are more important than similar proportional changes in R_o. Under conditions where resources are actually or nearly unlimited, then, fitness is raised by increasing birth rate (through its effect on R_o) and particularly by reducing the time between breeding periods. Hence in this sort of ecological setting one might expect the evolution of small (Section 5.1), fecund organisms (see also Meats, 1971).

As resources are progressively depleted the capacity for increase will become impaired. This can be represented most simply by the Verhulst-Pearl logistic equation:

$$\frac{dN}{dt} = rN \frac{(1-N)}{K} \qquad 7.4$$

where K = carrying capacity of the habitat or, in other words, the maximum number of organisms it will support. As N tends to K, growth rate reduces to zero and population density tends to a steady-state. Fitness here is not expressed in terms of growth rate, r, since there can be no population growth, but in terms of the extent to which populations (in inter-specific competition) or sub-populations (in intra-specific competition) can take a share of the limited resources; i.e. maximize their own K. Carrying capacity cannot be defined uniquely in terms of population traits since it depends on the interaction between the population and its environment (Stearns, 1977). This means that it is not possible to define precisely the population traits which should evolve under saturation conditions. Intuitively, though, it is clear that to maximize K, parents must not engage in activities that will lessen their own command over the limited resources and must produce offspring that are competitively superior. This is likely to mean a shift to an increase in the size of both the parents and the progeny (Section 5.1), a concomitant increase in developmental time (T) and, since reproduction is likely to put the parent at risk (see Section 7.3), to a reduced reproductive output.

The traits which promote fitness differ, therefore, depending on resource availability. The two extreme conditions have been referred to as 'r' and 'K' selection by MacArthur and Wilson (1967). Either pure 'r' or 'K' selection is probably rare in nature, however, and most habitats are likely to offer intermediate conditions which tend either to the 'r' end of the spectrum or the 'K' end of the spectrum (Pianka, 1970). Furthermore, Stearns (1977) has raised the question of the usefulness of the '$r-K$ approach' at all, since in practice it is difficult to take unambiguous predictions from it to natural populations with precisely defined dynamics. Nevertheless, the approach has given much stimulus to life cycle studies and for this reason I will sharpen the focus on it, on the theoretical ambiguities inherent to it, and on the practical difficulties associated with its application, in Sections 7.3 to 7.5.

7.2 Whether to reproduce all at once or repeatedly?

Cole (1954) distinguished between two sorts of organism – those that reproduce once and then die, and those which reproduce repeatedly. He referred to these respectively as semelparous and iteroparous strategies (Fig. 7.1). The question he asked was, 'under what ecological circumstances should a semelparous species become iteroparous? To do this he

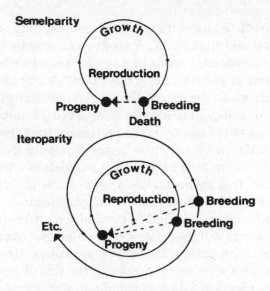

Fig. 7.1 Semelparous (upper figure) and iteroparous (lower figure) life cycles.

used r as a measure of fitness and examined two idealized situations. In his 'ideal semelparous population' the parents reproduced and died but there was no mortality between breeding seasons. Hence:

$$N_t = N_o (B_s) \qquad 7.5$$

where, N_t = size of population in one generation, N_o = size of population in preceding generation, B_s = average births/head (it is assumed for simplicity that we are dealing with a parthenogenetic population). In his 'ideal iteroparous population' all organisms were immortal and:

$$N_t = B_i N_o + N_o$$

so
$$N_t = N_o (B_i + 1) \qquad 7.6$$

If the iteroparous condition evolved from the semelparous then it can be assumed that $B_i = B_s$. Therefore the semelparous species could have achieved the same reproductive potential as the iteroparous simply by increasing its fecundity by 1. This now famous conclusion is referred to as 'Cole's result' and it raises the question as to why any species should have become iteroparous.

Cole's analysis was based, of course, on the implicit assumption that fitness is maximized by maximizing population growth. We have already seen that this need not be the case and if, as seems likely, there is some sort of trade-off between reproduction and adult survival (see Section 7.3) and the environment is such that the adult stands a better chance of surviving than progeny, then fitness may be maximized by reducing reproductive output and breeding repeatedly. Cole intentionally ignored the possibility of age-specific mortality but Charnov and Schaffer (1973) showed that if it is taken into account Cole's result can be modified drastically. Let the probability of survival of offspring to maturity be C and the probability of survival of the iteroparous parent from breeding season to breeding season be P. The population equation becomes:

for the semelparous population

$$N_t = N_0 . C . B_s$$

for the iteroparous population

$$N_t = N_0 . C . B_i + P . N_0 = (B_i C + P) N_0$$

and if the populations are to increase at the same rate:

$$B_s = B_i + P/C$$

Therefore, only if P and C are equal do we obtain Cole's result. If P is large and C is small, the probability of survival of offspring with respect to that of the parents is small and it would be necessary to add far more to the birth rate of the semelparous species to achieve the same fitness as the repeated breeder. Hence, under these circumstances the evolution of iteroparity would seem appropriate.

7.3 The cost of reproduction

The advantage of iteroparity is that it provides an insurance against the mass loss of progeny at any one time. In principle, therefore, the ideal life cycle strategy should bring together a high reproductive output and repeated breeding, since this would combine a high potential rate of increase with an insurance policy against the complete loss of progeny from one season's effort. In practice, however, such traits are found in association only rarely and in unusual circumstances (for example,

Life Cycles

Table 11 Approximate proportion of annual net assimilation involved in reproductive output. Data from Harper *et al.*, 1970.

Herbacious perennials	1–15%
Herbacious perennials including vegetative propagules and perennials without vegetative reproduction (e.g. trees)	5–25%
Wild annuals	15–30%
Most annual grain crops	25–35%
Maize and barley	35–40%

Table 12 Indices of reproductive output

$$\left(= \frac{\text{Egg vol.} \times \text{Egg no.}}{\text{vol. of Parent}} \right)$$

for several species of freshwater gastropod. (Data from Calow, 1978a).

Semelparous species	
Lymnaea pereger	1.23–4.88 (median = 3.05)
Physa fontinalis	2.17
Physa gyrina	2.01
Planorbis contortus	1.18
Ancylus fluviatilis	1.42–2.36 (median = 1.89)
Average *	2.06
Species with iteroparous populations	
Lymnaea stagnalis	0.34–0.51 (median = 0.43)
Lymnaea palustris	0.03–0.35 (median = 0.19)
Bithynia tentaculata	0.08–0.33 (median = 0.21)
Average*	0.28

* Based on median values

entoparasites often combine high fecundity with repeated, short-interval breeding; Jennings and Calow, 1975) and more often, we find inverse relationships between reproductive output and repeated breeding within taxa (e.g. Tables 11 and 12). The question now, therefore, is not why iteroparity should ever evolve, but why it has not evolved more frequently? In a sense this puzzle is the reverse of that posed by Cole, and the most obvious solution is that a premium has to be paid for the 'insurance policy' of repeated breeding. The simplest way of envisaging such a cost of reproduction is in terms of a negative causal relationship between the effort put into reproduction at one time and the subsequent reproductive value of the parent. Several models concerned with the evolution of life cycles make this assumption (Cody, 1966; Williams, 1966a and b; Gadgil and Bossert, 1970; Charnov and Krebs, 1973; León, 1976), and in this section we examine it more closely.

Negative correlations of the kind illustrated in Tables 11 and 12 are not sufficient in themselves to prove the existence of a particular kind of causal relationship. What is required is more evidence from carefully designed experiments and more specific information on the mechanisms that might be involved. For example, if there is a negative causal link between reproductive output and subsequent survival, it is crucial that gonadectomy or chastity should lengthen the life-span of semelparous organisms, and this has indeed been established for species as unrelated as tomatoes and fishes (Calow, 1977a). Furthermore, we know of several kinds of non-mutually exclusive mechanisms that might be involved in such a cause—effect chain. On the behavioural side, for example, it is clear that courtship, copulation, pregnancy, parturition and parental care can, by making parents more conspicuous or cumbersome, render them more prone to accident, disease or predation. For example, Tinkle (1969) has shown that short-lived lizards do indeed take more risks in these respects than longer-lived species. Alternatively, on the physiological side, it is obvious that if material and energy are used in reproduction (to produce gametes or to power activities associated with reproduction) when they are needed for other metabolic process then there will be serious, perhaps irreversible consequences (Calow, 1977a) on the survival chances of the parent. The excessive use of energy and resources in current reproduction might also impair future fecundity by weakening and perhaps even stunting the parent. Lawlor (1976), for example, has shown that by inhibiting the growth of *Armadillidium vulgare* (an isopod), reproductive output at one time has a negative effect on the future fecundity of this organism since fecundity is a size-dependent process. Relationships of this kind are also known to exist in the Plant Kingdom (Harper and White, 1974).

Reproduction, then, is a risky business and we need to consider when such risks are justified and when they are not. This is the evolutionary and ecological aspect of the problem. To carry out the analysis it is necessary to define two important parameters — *reproductive value* and *reproductive effort*. The former, reproductive value, is a term first coined by R.A. Fisher (1930) and represents the average number of young a female can expect to have over her whole life discounted back to the present. Rather conveniently it therefore takes into account both the chances of the female parent surviving to future breeding seasons, and her fecundity. Assuming steady-state conditions, with discontinuous breeding, the simplest possible mathematical formulation of reproductive value (v) is:

$$v_x = \sum_{t=x}^{max} P_t b_t \qquad\qquad 7.7$$

where: v_x = reproductive value of a female age x; b_t = fecundity (in female births) in each breeding season from t (including x) to *max*, the age when breeding stops; P_t is the probability of the female parent surviving from the current breeding season to the next breeding season. Obviously the equation becomes more complex for non-steady-state populations, because it is necessary in these to include a discount term for the relative contribution of offspring born late in life in a growing population, or early in life in a population that is decreasing in size, viz.:

$$v_x = \sum_{t=x}^{max} e^{-rt} P_t b_t \qquad\qquad 7.8$$

where: e = base of natural logarithms; r = coefficient of exponential growth (or shrinkage) of the population. This equation is very similar to Fisher's original formulation, though he was concerned with a continuous breeding system (integral replaces sum-sign) and measured v_x relative to the reproductive value of a female at birth (v_0). Note that by setting r to zero, e^{-rt} transforms to 1 and equation 7.8 reduces to equation 7.7.

It is now possible to partition both statements of reproductive value into two distinct components; one representing present reproductive output (b_x) and the other representing future potential reproductive performance, viz.:

$$v_x = b_x + \sum_{t=x}^{max} P_t b_t \qquad\qquad 7.9$$

or
$$v_x = b_x + \sum_{t=x}^{max} e^{-rt} P_t b_t \qquad\qquad 7.10$$

The second term on the right hand side of both equations is often referred to as the *residual reproductive value* (*RRV*) following Williams (1966b) and from the early discussion it is clear that this should be related in an inverse fashion to b_x. Any measure of fecundity which is proportional to its cost in terms of *RRV* is referred to as the reproductive effort (*RE*) of the parent. I will have more to say about RE in Section 7.5 but for the time being we take RE to be equal to b_x, which is quite sound as long as we are considering a single individual. It must be noted, however, that the nature of this trade-off between RE and RRV is likely to be important in determining life cycle pattern since the less sensitive RRV is to RE then the more feasible becomes repeated breeding (e.g. Schaffer, 1974). Unfortunately, no one has succeeded previously in defining this relationship and it may be complex (Stearns, 1977). In the analysis that follows I assume that the trade-off is the same for all cases.

Carrying on the spirit of the discussion in Section 1.5 and Section 7.1 it is not difficult to adjust equations 7.9 and 7.10 to give measures of fitness (F) under conditions of 'r' and 'K' selection. All that is necessary is to include terms for the survivorship of the offspring, for it is clear that fitness not only depends on the total numbers of offspring that are produced, but on the numbers produced that are themselves capable of going on to reproduce. Let C_x = probability of survivorship of individuals from birth to maturity. Then:

$$F_{(r)} = C_x b_x + \sum_{t=x}^{max} e^{-rt} C_t P_t b_t \qquad\qquad 7.11$$

$$F_{(K)} = C_x b_x + \sum_{t=x}^{max} C_t P_t b_t \qquad\qquad 7.12$$

These equations give some insight into how much effort individuals should put into reproduction at any one time, and about the ecological distribution of life cycle strategies. Firstly, we see that what is important from the point of view of fitness is not the reproductive performance of the organism at one time, but its life-time reproductive performance. As a general evolutionary rule it is clear, therefore, that current RE should only be increased as long as the gains so achieved are greater than the concomitant reduction in RRV. Second, since both b_t and P_t are likely to reduce with age then so is RRV, and this predicts that RE should increase with age — a prediction which has been corroborated in birds (Stearns, 1976) and lizards (Pianka and Parker, 1975). Note, however, that because of the exponential term in equation 7.11, RRV is likely to reduce more rapidly under 'r' conditions of selection, provided, of course, that all other parameters are equal. This, together with the fact that density-independent mortality factors often operate in an age-independent fashion such that $C = P$, means that even more dramatic reductions in RRV will occur with age under 'r' selection as compared with 'K' selection.

Alternatively, density-dependent factors, like competition and predation, are likely to strike the pre-reproductive individuals more than the immature individuals such that $C \ll P$. This would tend to predict, therefore, that iteroparity should be a more common feature of 'K' selected organisms, and that semelparity should be a more common feature of 'r' selected organisms (see Section 7.2). It is not beyond the bounds of feasibility, however, that certain kinds of 'r' selection might disadvantage young relative to old and, though less conceivable, certain kinds of 'K' selection might disadvantage old relative to young. In these cases the initial predictions would be reversed. To make specific predictions, therefore, it is necessary to investigate the effects of specific demographic characteristics on equations 7.11 and 7.12 and computer simulation experiments using this kind of approach have been carried out by Murphy (1968) and Gadgil and Bossert (1970). Since each of these studies begins with different demographic premises with respect to the susceptibility of the old and young, it is not surprising that they end up with different conclusions; Gadgil and Bossert favouring the prediction that 'r' selection results in semelparity and 'K' in iteroparity, but Murphy favouring the opposite. An alternative method is to match life cycle patterns with the demographic properties of populations in real ecological circumstances, and in the next section I consider a few examples of this kind of approach.

7.4 Examples of reproductive strategies

In conditions of intense competition for limited resources, it is likely that a large 'experienced' parent will fare better than small offspring. Alternatively, in circumstances of intense density-independent (D.I.) mortality, a high intrinsic rate of increase and thus a high reproductive effort will be necessary to counteract the chances of decimation. It is to be anticipated, therefore, that reproductive effort should be greatest when mortality is erratic and least when density-dependent factors predominate — this is the most obvious prediction in the life cycle consequences of 'r' and 'K' selection discussed in the last section. It has also been advocated by Rabinovich (1974) and Southwood (1976).

Abrahamson and Gadgil (1973) have tested this hypothesis with observation on herbs of the genus *Solidago* (Compositae). As an index of reproductive effort they used the ratio of floral biomass to total biomass, each expressed in energy units. Six populations of four species (*Solidago canadensis*, *Solidago rugosa*, *Solidago speciosa* and *Solidago nemoralis*) were considered in three separate communities. One community

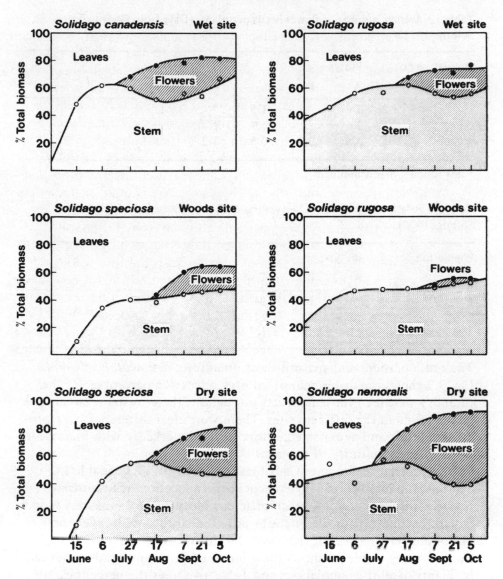

Fig. 7.2 Percentage total biomass of stem, leaves and flowers as a function of time in *Solidago* spp. (With permission from Abrahamson and Gadgil, 1973, *Am. Nat.* **107**. Copyright by University of Chicago Press.)

labelled *dry* was dry and heavily disturbed (i.e. high D.I. mortality). The second, labelled *wet* was a wet meadow site and the third, labelled *hardwood* was dominated by trees of the genus *Betula*, *Quercus* and *Acer*. The latter was least disturbed and was judged to have the least D.I. mortality.

Table 13 Average number of flower heads per plant. (Data from Gadgil and
Solbrig, 1972).

Population no.	Biotype			
	A	B	C	D
1	3.6	2.3	1.5	*
2	2.6	2.1	1.9	†
3	3.8	2.3	0.5	1.2

* not found   † low numbers

Table 14 Relative abundance of biotypes at each site. (Data from Gadgil and
Solbrig, 1972).

Population no.	Biotype			
	A	B	C	D
1	73	13	14	0
2	53	32	14	1
3	17	8	11	64

The order of increasing maturity was, therefore: *dry, wet, hardwood.*
Fig. 7.2 shows the dry weight of inflorescences as a percentage of the
total dry weight of the aerial parts throughout the growing season in the
plants found in the different sites. There were clear differences in reproductive effort and as expected, effort reduced markedly with increase in
the ecological maturity of the site.

Using similar techniques, Gadgil and Solbrig (1972) were able to
demonstrate intraspecific differences in the allocation of resources
between the stem and flowers of different biotypes of dandelions found
in habitats differing in the intensity of D.I. disturbance. Isozyme and
morphological analyses of three populations revealed the existence of
four biotypes. Table 13 shows the number of flowering heads produced
by plants of all the populations and Table 14 shows the percentage of
each of the four dandelion biotypes in each of the three populations. The
degree of D.I. disturbance was judged to reduce from population 1 to
population 3. As expected, biotype A, which put the most effort into
reproduction, was most abundant in population 1 — the one most subject
to D.I. mortality. Alternatively, biotype D, which put the least effort
into forming flowers, was most abundant in population 3 — the one with
least D.I. mortality. Other biotypes were intermediate with regard to

Table 15 Approximate egg production (per individual per season) in several species of woodland millipede. (Data from Blower, 1969).

Species	Age (years)					
	1	2	3	4	5	6
C. latestriatus *	0	0	10	10	12	0
C. punctatus *	0	0	0	14	40	130
O. pilosus †	0	0	60	–	–	–
I. scandinavius †	0	0	0	150	–	–

* Iteroparous   † Semelparous

reproductive effort and also with regard to relative abundance in each of the three populations. Other studies on 'r' and 'K' strategies in plants are those of Hickman (1975) and McNaughton (1975).

In contrast to the above predictions it has been suggested that when conditions are unpredictable, parents should 'hedge their bets' and should produce fewer offspring, thereby conserving themselves (e.g. Murphy, 1968). 'Bet-hedging' is only appropriate, however, when the parent has a better chance of surviving than all the offspring that have been produced to 'save' the parent. Whenever the chances of survival of progeny are continuously low it is nevertheless advantageous for adults to carry out replacement over a fairly long period of time. Under normal circumstances this requires a reduced reproductive effort at any one time and repeated breeding.

Low replacement and high juvenile mortality may occur when most of the suitable sites for establishment are filled or when they are few and far between as might occur in colonizing species which in every other respect are 'r' selected. Blower's (1969) work on millipedes illustrates how the patchy distribution of preferred habitat might influence life-history. In the millipedes two species, *Cylindroiulus punctatus* and *C. latestriatus,* are iteroparous whereas two other species, *Iulus scandinavius* and *Ophyiulus pilosus,* are semelparous. On average the semelparous species put more effort into reproduction in a season than the iteroparous species (Table 15). Each group has a slightly different ecology. The two semelparous species are evenly distributed amongst the soil and litter of deciduous woodland but the two iteroparous species have a very patchy distribution and this may be associated with an apparent preference for dead logs. From this Blower (1969) conjectures that repeated breeding and thus reduced reproductive effort is an adaptation which increases the chance of at least one offspring finding a new log. Entoparasites have exactly the same problem in locating new hosts and deal with it in a similar way. Here, life in a rich, protected habitat (the host) allows the parent to breed continuously at a high rate with little risk (Jennings and Calow, 1975).

Not all studies have produced results which fit so neatly into a theoretical framework as those above. Out of 35 good studies on organisms ranging from insects to mammals Stearns (1977) was able to identify 18 which conformed to theory and 17 which did not. The poor fit of the 17 could have been attributable to a host of factors. For example: in any natural population there may be a system of multiple causation (e.g. a mixture of 'r' and 'K' influences – Jennings and Calow, 1975) which works out a compromise rather than a clean effect; biological constraints may prevent the predicted optima from ever being achieved; short-term, proximate, ecological influence may modify or even simulate (Ricklefs, 1977) the effects of long-term selection. Alternatively, the practice may be at fault since it is not always possible to define how populations are being regulated, nor to quantify parameters like reproductive effort (Section 7.5). This means that in attacking the life cycle problem, much care has to be given to the choice of material and the observations we make on it. Finally, it has been suggested that the theory itself might be inadequate; for example, since it rests on a view of fitness which plays down sex (Section 7.1). Intuitively, though, it would seem likely that such short-comings are more likely to influence the details of the predictions rather than their general form. The problem is complex and the approach naive but some progress has nevertheless been made in an understanding of when organisms should reproduce, and by how much.

7.5 Reproductive adaptability

The effort put into reproduction is adapted to meet a particular sort of environmental challenge. In the life of one individual the intensity of the challenge may change and it is necessary to consider to what extent this might influence the reproductive output. It goes without saying that reproductive adaptability is likely to be adaptive.

In many respects it is profitable to consider the organism as a device which partitions input energy (I) into that needed to produce gametes (*Rep*) and that needed for other aspects of metabolism (*Rest*). *Rep/I* is therefore the transfer function of the system (Calow, 1976) and is some-times used as an index of reproductive effort (see Section 7.3). The

rationale behind this is that the more energy that is used up in *Rep*, the
less that will be available for *Rest* and, as noted in Section 7.3, this is
likely to reduce the future reproductive performance of the parent and/or
the chances of it surviving after reproduction. Certainly, from the point
of view of comparisons between individuals of the same and different
species, *Rep/I* is superior to measures of reproductive effort expressed in
terms of the number of gametes per parent, since the latter fail to take
into account that both the size of the gametes and the parent can show
considerable variation, particularly from an interspecific point of view.
It is quite conceivable, for example, that a small individual that produces
a small number of very large gametes is putting more effort into reproduction than a large individual which produces a large number of very
small gametes (Hirshfield and Tinkle, 1975). On the other hand, it is as
well to remember that selection operates on the basis of the availability
of energy for growth and reproduction, not necessarily on the basis of
the conversion efficiency, though this might become important in the
later stages of evolution (Section 2.3). Similarly, for a given size and set of
conditions, maintenance metabolism is likely to require a fixed amount,
not a fixed proportion, of the energy input; and any failure to supply this
amount is likely to have serious consequences on the well-being of the
parent.

Consider now what might happen to the partitioning of energy between
Rep and *Rest* if *I* becomes reduced — probably a common occurrence in
the breeding season as more mouths become recruited into the population
and as parents often have less time to spend on feeding. Under such conditions reproductive effort, the influence of the act of reproduction on
the future reproductive value of the parent, remains constant if, and only
if, the flow of energy into *Rest* remains unimpaired. If *Rest* remains
constant but *I* becomes reduced *Rep/I* must also become reduced. This
is illustrated in Fig. 7.3a and b. If, alternatively, *Rep/I* remains constant
despite reductions in *I*, *Rest* must reduce in proportion to *I* and in this
case the system cannot be said to be maintaining a constant effort. This
is illustrated in Fig. 7.4a and b. Hence, in comparing individuals and
populations at different levels of ration, *Rep/I* provides only a crude
index of effort. A better, but more difficult to measure index would be
$[\![1 - [(I-Rep)/Rest^*]]\!]$; where $Rest^*$ is the energy used in *Rest* under
optimal conditions. Call this index RE^*.

Consider again Fig. 7.3b. The curve shows the way *Rep/I* would have
to alter if reproductive effort remained constant (RE^* remains at 1).
I refer to organisms that employ this strategy as 'conformers'. Organisms

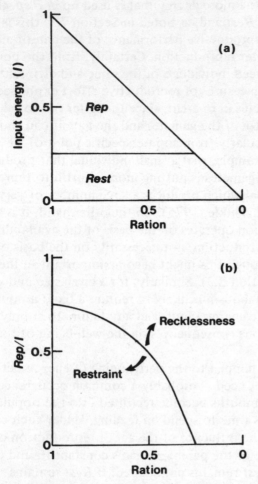

Fig. 7.3 Graph (a) illustrates how input energy (*I*) is allocated to reproduction
(*Rep*) and other elements of metabolism (*Rest*) at various ration levels.
Graph (b) shows how this influences the efficiency ratio *Rep/I*. See text for
further explanation.

using energy allocation strategies that fall below the curve can be said to
be showing reproductive restraint, since in these cases reproductive effort
falls proportionately more than the reduction in ration ($RE^* < 1$). Here
the effect of reproduction on the parent becomes more intense under
good feeding conditions. In desert rodents, for example, low survival
rates and high levels of reproduction are associated with years of good
plant production while high survival and low reproductive rates are
associated with years of poor plant production (Nichols *et al.*, 1976).

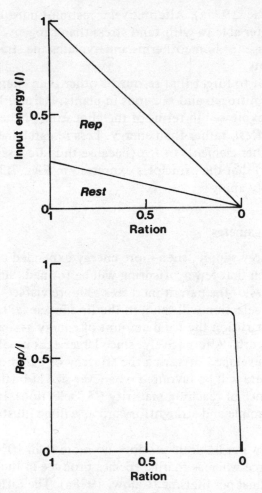

Fig. 7.4 As in Fig. 7.2 but for a 'reckless reproducer'. See text for further explanation.

Alternatively, species with allocation strategies falling above the curve in Fig. 7.3b may be said to be showing reproductive recklessness in that reproductive effort is negatively related with ration ($RE^* > 1$). This kind of strategy seems to occur in some snails and triclads (Calow and Woollhead 1977a; Calow, 1978a) where Rep/I either remains constant (as in Fig. 7.4) or rises under conditions of reduced ration.

In general evolutionary terms it is likely that the adaptability of reproductive strategies is itself adaptive. Recklessness will only be selected when progeny have an equal or better chance of surviving through harsh nutritive conditions than parents. This seems to be the case in some triclads (Calow and Woollhead, 1977a). Alternatively, restraint must be shown when parents are better able to withstand stress than progeny. This is probably the usual case for homeotherms and explains the strategy observed in the desert rodents.

Finally, it is important not to forget that resources other than energy may be in short supply (e.g. nitrogen and minerals in plants; Harper, 1977) and that *RE* may be better expressed in terms of the allocation of these resources between *Rep* and *Rest,* rather than energy. Less research has been carried out on these other elements of *RE* (because thus are less easily measured than energy) that the principles expressed in Figs. 7.3 and 7.4 will also undoubtedly apply.

7.6 The size and number of gametes

If a female has a limited energy supply, then more energy expended on individual offspring will mean that fewer offspring will be formed. Since, under '*r*' selection, the fitness of the parent increases as more viable progeny are produced, then selection will tend, in the first instance, to favour those species which partition the total amount of energy available for reproduction into most parts. Alternatively, since larger eggs usually result in larger, more fully developed offspring the strategy of partitioning energy into a few, large quanta will be favoured whenever size at birth is correlated with a better chance of reaching maturity ('*K*' selection). However, thing s are rarely this simple and straightforward as will be illustrated in the examples that follow.

Gastropods living in freshwater habitats produce no more than 10^3 eggs per individual per lifetime whereas marine species produce in the order of 10^6 eggs per individual per lifetime (Calow, 1978a). The difference is correlated with the fact that as gastropods have invaded more inclement freshwater environments, the larval stages, which in marine forms are usually planktonic, have been retained increasingly in the protective envelope of the egg. It has been necessary, therefore, to increase the space and provisions allowed to the developing embryo and thus to concentrate on egg size rather than numbers (Calow, 1978a).

In commensal and parasitic copepods, large eggs are associated with stable habitats where hosts are abundant, whereas small eggs are produced by species which inhabit unstable (exposed) habitats where hosts are few and far between. According to Gotto (1962) the rationale behind this is that large larvae from large eggs have a competitive advantage in safurated situations whereas large numbers of larvae will be needed in exposed habitats, where there is likely to be high larval mortality. This

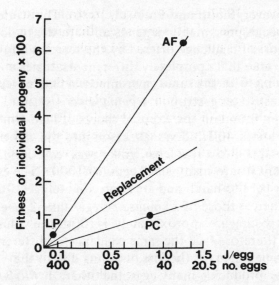

Fig. 7.5 Smith and Fretwell's (1975) fitness set analysis on the reproductive strategies of three species of freshwater gastropods (AF = *Ancylus fluviatilis*; LP = *Lymnaea pereger*; PC = *Planorbis contortus*) living in harsh, littoral habitats. Straight lines on the graph represent lines of equal fitness. The line marked *Replacement* is the strategy that must be adopted if there is to be no net reduction in population density with time. *L. pereger* and *A. fluviatilis* are discussed in the text. *P. contortus*, though able to live in littoral habitats, is not able to meet the replacement requirement. Populations of this species are maintained by immigration from more sheltered refuges. (With permission from Calow, 1978a).

last state of affairs is probably typical of most entoparasites where larval death during transmission is likely to be great. Here, though, eggs are often numerous and large due to the very rich conditions in which the parent parasite finds itself (Jennings and Calow, 1975).

The problem of seed design in plants is also complex (Harper *et al.*, 1970). Most plants exist in a complex, co-evolutionary context contending with agents of destruction, like fungi, viruses and predators and the requirements of disperal. Optimizing the size specification of seeds for one purpose need not optimize them for another. Small seeds are favourable for wind dispersal and for avoiding insect predators. Alternatively, large seeds with thick coats escape many predators and have sufficient resources to support rapid, initial growth, which would be important in situations of intense competition.

It is difficult to come to any specific conclusions, therefore, about egg

size and numbers (see, however, Smith and Fretwell, 1974). The strategy
adopted by a particular species presumably depends on both the biology of
the species itself and the sort of challenge offered by the surrounding environment. It may even be true that completely different strategies are
equally satisfactory in dealing with the same environmental challenges. For
example, two species of freshwater gastropod, *Lymnaea pereger* and
Ancylus fluviatilis, are often found on the exposed shores of lacustrine
habitats. Both species put about 40 J/individual/season into the gametes
but *A. fluviatilis* partitions this into a few large, yolky eggs (*ca* 30/individual)
whereas *L. pereger* partitions it into many small eggs (*ca* 1000). The eggs
of *A. fluviatilis* take four weeks to hatch and are approximately 1/10th
the volume of the adult whereas those of *Lymnaea pereger* take 2 weeks
to hatch and the emergent young are approximately 1/100th the volume
of the adult. Both species therefore face the same challenge in different
ways. One, *L. pereger,* compensates for the loss of young due to the
scouring action of waves by producing many eggs; the other, *A. fluviatilis,*
protects the young by ensuring that they are bigger and thus better able
to 'hang on' in the face of scouring. Taking into account the subsequent
survival of hatchlings to maturity, it is possible to show that these
divergent strategies are equally fit under the same conditions (Fig. 7.5).
This may be a common feature of complex adaptations like life-history,
and perhaps feeding strategies.

Part four Ageing

Life and death are inextricably linked. All living things ultimately fall
foul of accident, disease or predation. Occasionally, however, we observe
them wearing away or dying naturally, apparently as a result of some
intrinsic, degenerative process. In the following chapters I turn to this
degenerative part of life's cycle.

There are several very important questions that can be asked about
ageing; What is it? What causes it? Has it any evolutionary significance?
Each of these questions is considered in Chapter 8 where I develop a
general theory which brings together both the immediate physiological
changes associated with ageing in the whole organism, and the ultimate
evolutionary forces which lie behind its occurrence. Finally, in Chapter 9,
I investigate processes which are apparently able to rejuvenate ageing
systems and which, in principle, are able to confer immortality. These
give further insight into the ageing process.

8 The ageing process

8.1 What happens to organisms which do not meet an unnatural death?

One way of studying mortality is to record the age at death in a population of similar individuals. Thus it is possible to plot out survivorship curves for a group of organisms all born at the same (a cohort). For example, curve I of Fig. 8.1 shows a survivorship curve for a natural population of ramshorn snails. Here the probability of death reduced with age, and this can be ascribed to the fact that in nature small, young snails are more often susceptible to predation, disease or accident than are older organisms. Sometimes these agents of 'unnatural' death act in an age-independent manner as is shown for another species of snail in curve II of Fig. 8.1. Take 'unnatural' death away, however, as in domestication and a welfare state, and there is still mortality — a mortality which strikes the old more than the young. This sort of process is illustrated in curve III of Fig. 8.1 for the ramshorn snail under laboratory conditions, which provided plenty of food but excluded predation. If the possibility that curve III was caused by a gradual deterioration of the environment can be excluded, then it must have been brought about by deterioration intrinsic to the organisms themselves. This is the process of ageing or senescence. The survivorship curve for the human population has probably shifted from a Type I–II curve in ancient times to a Type III curve in recent times. This has been associated with a marked improvement in the conditions of life brought about by improvements in hygiene and medical care (Strehler, 1962).

In a cohort of ageing individuals the change in numbers with time occurs according to the Gompertz equation (Strehler, 1962):

$$dN/dt = -R_m N \qquad 8.1$$

and

$$R_m = R_o e^{\alpha t} \qquad 8.2$$

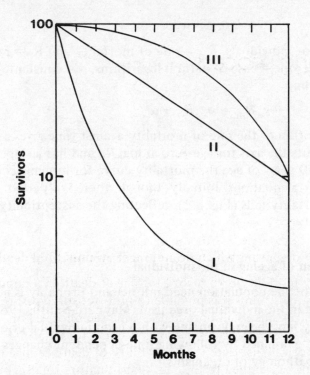

Fig. 8.1 Survivorship curves of *Planorbis contortus* in the field (curve I) and the laboratory (curve III), and survivorship curve of *A. fluviatilis* in the field (curve II). Data for curves I and II are from Calow (1978a). The time scale should be multiplied by about 2.5 for curve III.

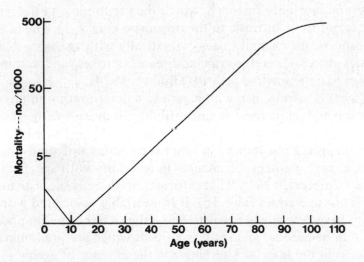

Fig. 8.2 Approximate Gompertz plots for Western European males.

where N = numbers of individuals, R_m = rate of mortality and R_o = rate of mortality at age zero, e = base of natural logarithms, α = constant, t = time. It follows that:

$$\log_e R_m = \log_e R_o + \alpha t \qquad\qquad 8.3$$

So plotting the logarithm of the rate of mortality against time gives a straight line which cuts the axis for age zero at $\log_e R_o$ and has a slope of α. Between 20 and 80 years of age the mortality curve for humans conforms to the Gompertz equations. Initially, though, there is a period when the rate of mortality falls (Fig. 8.2), reflecting the susceptibility of infants to mortality.

8.2 The manifestation of ageing in the individual

The mortality curve of the population need not be, and probably is not, a reflection of ageing in the individual organism (Maynard-Smith, 1962). Ageing processes may have been occurring within each individual prior to death. This can be examined by following time-dependent changes in the physiology and pathology of organisms.

In man, after the age of thirty, there is a steady decline of about 1 to 3% per year, in the maximum capacity of many physiological functions; for example in cardiac index, standard glomerular filtration, maximum breathing capacity and so on (Kohn, 1971a). At the same time, we become more susceptible to death from various infectious diseases and from several mortality factors in which the predominant etiological component seems to be intrinsic to the organisms (Fig. 8.3). The incidence of all cancers, for example, rises dramatically with age as do deaths from coronary diseases, cerebrovascular diseases (strokes) and autoimmune diseases like rheumatoid arthritis (Burnet, 1974).

Much of what is seen in man with regard to a deterioration in physiological function and an increase in susceptibility to disease with age also seems to occur in mice and rats (Simms and Berg, 1957). There is less information on ageing phenomena in other vertebrates and almost no information, except, perhaps, on changes in fecundity with age, for the invertebrates (Comfort, 1964). What information there is suggests that senescence is not universal (Table 16). It is invariably associated with groups which have determinate growth and either a fixed cell number per individual, as in nematodes and rotifers, or cells which are predominately post-mitotic, as in the insects. Alternatively, the absence of ageing phenomena is usually associated with organisms which have indeterminate

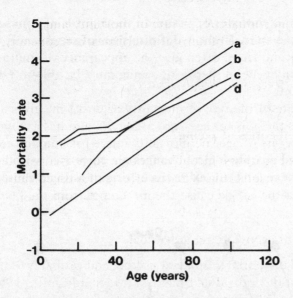

Fig. 8.3 Approximate Gompertz plots of cause-specific mortality rates.
(a) total causes; (b) diseases of the heart; (c) other diseases; (d) diseases of
CNS and vascular systems. Mortality rate units are powers of 10. (Data from
Strehler and Mildvan, 1960).

Table 16 Distribution of senescence between phyla in the Animal Kingdom. (Data
collected from Comfort, 1956 and Strehler, 1962).

	Suspected lack of ageing	Suspected ageing in some	Definite ageing in some
Coelenterates	✓	✓	
Platyhelminthes	✓	✓	
Molluscs	✓	✓	
Nematodes			✓
Annelids	✓	✓	
Rotifers			✓
Arthropods			✓
Fishes	✓		✓
Amphibia	✓		✓
Reptiles	✓	✓	
Birds			✓
Mammals			✓

growth and a continuous turnover of tissue. In the Plant Kingdom some species, in particular the annuals, show definite signs of senescence, whereas others, notably shrubs and trees, often give the appearance of immortality (Woolhouse, 1972). Any general theory of ageing must be able to take these facts into account.

8.3 Single or multiple theories of ageing?

Many physiological and morphological changes are correlated with ageing. Are these the effects of several causes or the effects of a single cause (Fig. 8.4)? Support for the 'single cause theory' comes from the fact that

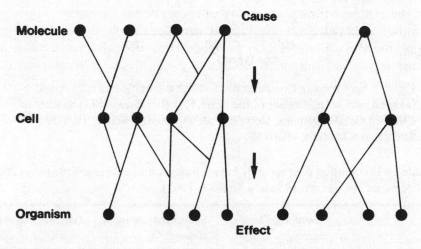

Fig. 8.4 The way in which several effects can be produced by one or many causes.

all the physiological changes associated with ageing are closely synchronized in temporal expression. However, Maynard-Smith (1962) has argued that this synchrony might occur through the synchronization of disparate causes as a result of the action of natural selection. Any mechanism which caused a too rapid deterioration in some character would be selected against and this would ultimately lead to complete synchronization.

There are two pieces of evidence which support the theory of multiple causes. First, heterochronic transplantation procedures in mammals have shown that within limits some organs retain the age characteristics of the donor when transferred to either young or old hosts (Krohn, 1962, 1966; Hollander, 1970). This suggests that organismic parts age at a rate determined by intrinsic properties rather than by extrinsic, organismic properties. Second, senescent deterioration, weakening and ultimate death may occur in one of several ways.

On an interspecific and interphyletic level there are also good grounds for believing that there is more than one mechanism behind ageing. As noted in the previous section, some organisms appear to be immortal, and in others ageing phenomena may be reversible (Chapter 9). Ageing and mortality in some species are strongly associated with reproduction (Section 7.3). Nematodes, rotifers and insects with a predominance of post-mitotic tissue in the adult, age by irreversible degeneration of irreplaceable parts of the body. Alternatively, most vertebrates have mitotically active tissues, some of which continue to turn over through-out life, and yet they still age.

The evidence for a multiplicity of causes behind senescence, both within and between species, is persuasive yet the physico-chemical basis of ageing may still depend on a single sort of process. Cars may age in many ways and from many causes and yet as Maynard-Smith pointed out (Maynard-Smith, 1962) these are usually associated with just a few under-lying processes — abrasion, corrosion and metal fatigue. We investigate the possibility of finding a unitary basis for the apparent multiplicity of causes behind ageing in Section 8.5.

8.4 Ageing by accident or design?

Sooner or later, in the study of ageing and death, it becomes necessary to consider whether an ageing process is built explicitly into the design of living things or whether ageing is simply due to the random accumulation of damage within the system.

The predictability of ageing phenomena both within and between species gives no doubt that they are part of a programmed process. Furthermore, the good correlation between the life-span of parents and offspring, and the similarity in the life-span of twins reared apart points strongly to a genetic basis to ageing (Korenchevsky, 1961; Strong, 1968). However, this need not means that suicide instructions are written explicitly into the genetic programme. An alternative could be that organisms age as a result of the way that they are designed to do other things. Since, in physiological and pathological terms, deterioration is always likely to be attributable to the way the system is designed and built irrespective of whether such an effect was *intended* by selection, it is not permissable to argue from the design-like progression of senescence in the individual to the idea that there has been positive selection for it.

This question can only be decided on the basis of evolutionary logic.

Are there any circumstances in which there might be a positive selection pressure for senescence and death? In terms of sheer evolutionary logic, anything which reduces total potential fecundity, as decay and death must do, will have a negative, not a positive effect on fitness. Alternatively, it has often been argued (e.g. Weismann, 1891) that death is needed in the evolution of life to make room for new adaptations and to prevent parents too decrepit for reproduction from coming into competition with more viable offspring. However, this point of view is associated with a number of difficulties. First, as Medawar (1952) first noted, the argument is viciously circular, assuming what it has set out to prove, in particular that old animals are too decrepit to reproduce. Secondly, it is not clear how important this 'adaptation' might be since in nature most organisms die 'unnaturally' by disease, predation or accident, so the chances of an adult living very long, even in the absence of senescence, are never very great. Finally, even if important, it is not clear that selection could favour an adaptation that operates on an inter-generation basis. The problem here is that the gene for dying would not only give advantage to progeny carrying it, but to conspecifies without the gene. This is because the death of the adult would lead to a general alleviation of local competition. Therefore, apart from special circumstance, as envisaged in kin selection (see Section 1.5 and also Maynard-Smith, 1964), the gene for dying could not be selected for.

The weight of evidence seems, therefore, to support the hypothesis that ageing is not designed explicitly into the system, but has evolved as a side-effect of other characters which have a definite, positive, selection value. Several specific evolutionary theories are based on this conclusion; namely those of Medawar (1952), Hamilton (1966), Williams (1957), and Edney and Gill (1968). Though these differ in detail they are all founded on the same basic logic. The points common to them all are as follows:

(1) the older an organism becomes the more chances it has of dying unnaturally;
(2) the probability that any organism will leave offspring (*RRV*, see Section 7.3) reduces with age and there comes a point when, under normal circumstances, most organisms are dead;
(3) beyond this point in the life cycle there is a selection shadow (Calow, 1977a) in which the genes escape the scrutiny of selection;
(4) it is demonstrable, therefore, that genes which confer a reproductive advantage early in life will be selected for even if they have ill-effects later;

(5) under conditions where organisms are protected against 'unnatural death', as in the laboratory and the Welfare State, these ill-effects will be expressed;
(6) this is the basis of senescence.

According to these theories, senescent changes ought to represent a mixed bag of side-effects from any genes that happen to have favourable effects in younger stages. As already noted, there is evidence for multiple causes behind ageing, but the process is sufficiently predictable to point to the involvement of a single type of process. Furthermore, there is some circularity in the Medawar-Williams hypothesis, for temporal changes in gene expression are dependent on temporal changes in the molecular and cellular environment of the genes and hence on an ageing process. The next section, therefore, attempts to solve these problems by suggesting a mechanism of ageing based fundamentally on general molecular deterioration which takes its cues from ordinary developmental changes.

8.5 A hypothesis based on the random, non-programmed accumulation of molecular damage

Many biochemical molecules, if preserved in an inert state, can retain their biochemical potency for years (Abelson, 1957; Keilin, 1953). Even some whole organisms can be maintained in a state of suspended animation at extremely low temperatures (Hinton, 1968). However, should frozen molecules and organisms be described as alive just because they are able to store the blue-print for life processes? From the two preceding Parts of the book, life would seem, above all, to be a cyclical, metabolic activity and according to the argument proposed in Chapter 1, it is through such physiological activity that living systems and their parts are likely to become damaged and disorganized.

Several specific mechanisms may cause molecular damage and in turn, several ageing theories are built upon these processes. The three most common theories involve DNA errors (Curtis, 1971; Burnet, 1974), errors in protein synthesis (Orgel, 1973) and general molecular damage through the formation of cross-links (Cutler, 1975).

(1) *The somatic mutation theory of ageing arose mainly out of the work of Howard Curtis.* Errors in the DNA of somatic cells are supposed to accumulate with age, eventually leading to a loss in the normal biological function of the cells in which they occur (Curtis, 1971). Curtis based his theory on the observation that the frequency of chromosomal aberrations

increased with age and that X-irradiation caused similar aberrations and brought about a shortening of life-span. There are, however, several inconsistencies in this theory (Curtis and Tilley, 1971; Goldstein, 1971; Price and Makinodan, 1973). For example, the correlations between radiation type, dose effects, chromosomal aberration and life-shortening are imperfect and the relationship between DNA errors and gross chromosomal damage has never been fully established. Several more complex theories have been put forward to reconcile some of these difficulties, and the most notable are the ones which propose that the accumulation of somatic mutations is enhanced by a reduction in the fidelity of DNA replication and repair (Burnet, 1974; Little, 1976). Mammalian cells are known to possess enzymic machinery for recognizing and repairing DNA damage (Section 4.2) and there is good evidence for an age-associated decline in the fidelity of these repair mechanisms (Wheeler and Lett, 1974; Schweitzer and Bodenstein, 1975; Little, 1976; Linn *et al.*, 1976). For example, Epstein *et al.* (1974) and Mattern and Cerutti (1975) have reported that the rate of DNA repair is reduced *in vitro* in human senescent fibroblasts and progeria, and the synthesis of DNA by a low molecular weight DNA polymerase shows more errors when extracted from old animals (Barton and Young, 1975; Bolla and Brot, 1975).

Some suggest that gene repetition will protect against damage and that in consequence the life-span of an organism will depend on the extent to which its genetic messages are repeated (Medvedev, 1972). It may also be true, however, that in circumstances of repetition, life-span will depend critically on damage to any little-repeated or non-repeated genes. Hence, it is not surprising to find that various genetic diseases apparently effecting the acceleration of ageing, for example Werner's syndrome, are dependent on the mutation of a single gene.

(2) *The theory that cellular ageing is related to the accumulation of errors in protein synthesis has been formulated most concisely and most provocatively by Orgel (1973).* He suggested that if errors turn up in the transcription—translation machinery of DNA they could become self-perpetuating, leading to a gradual but irreversible and accelerating breakdown in the fidelity of protein synthesis, and ultimately leading to an error-catastrophe. One of the major predictions from this theory is that the enzyme population within an organism should change with age away from optimal functioning. This has been demonstrated experimentally in certain mammalian and non-mammalian systems (Gershon and Gershon, 1973; Holliday and Tarrant, 1972; Wulf Cutler, 1974; Goldstein *et al.*, 1975). Furthermore, premature senescence has been

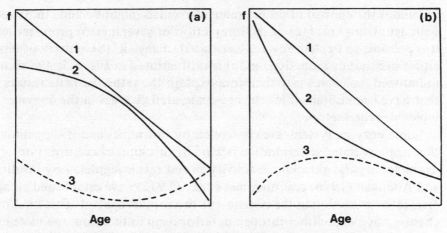

Fig. 8.5 Theoretical curves showing how changes in enzyme concentration (1) and enzyme inhibitor concentration (2) with age can bring about different kinds of change in enzyme activity (3). f = activity level. (With permission from Hall, 1976, *The Ageing of Connective Tissue,* Academic Press, London and New York. Copyright by Academic Press Inc. (London) Ltd.)

induced by growing organisms in the presence of base analogues known to cause errors in protein synthesis (Holliday and Tarrant, 1972; but see Bozouk, 1976), and abnormal enzymes have been isolated from the fibroblasts of individuals suffering from syndromes which induce premature ageing (Goldstein *et al.,* 1975). Alternatively, several pieces of important evidence contradict the theory since many enzyme systems do not appear to deteriorate with age (Baird *et al.,* 1975, but see Holliday, 1975 and the reply of Baird and Massie, 1975; also see Danot, Gershon and Gershon, 1975, Eisenbach *et al.,* 1976; Yagil 1976; Rubinson *et al.,* 1976). Also the growth of human cells in the presence of base analogues did not lead to premature ageing *in vitro* (Ryan *et al.,* 1974) and polio or herpes viruses produced from senescent cells were no different in morphology or in protein structure to those produced from young cells (Holland *et al.,* 1973; Tomkins *et al.,* 1974). These results suggest that the protein-translating machinery involved in enzyme production remains intact in aged cells. However, they do not exclude the possibility that other, error-containing proteins (perhaps concerned with cell structure) were accumulating.

As well as a direct effect on enzyme activity, there is also a well-established deterioration with age in enzyme induction (Adelman, 1975; Eisenbach, *et al.,* 1976). Inducible enzymes may be affected in terms of magnitude or time-course of response, and it is likely that these properties are under the control of other molecules which might become impaired with age. Using this idea of the interaction of several error-prone molecules it is possible to explain any age-associated change in the activity of enzymes, either increase or reduction, and this is illustrated in Fig. 8.5. Age changes in induced molecules may therefore explain the rather variable results that have been obtained for the age-associated changes in the enzyme molecules themselves.

Some enzyme systems are controlled by hormones and it is possible that age-associated deterioration occurs in this kind of control. For example, hepatic glucokinase activity in fed rats is regulated by insulin and Adelman (1970) and Adelman *et al.*, (1972) have established an age-associated reduction in the potency of the insulin control. This kind of change may occur either through deterioration in the hormone molecule itself or in the molecular make-up of the hormone receptor sites on the cell surface (Adelman, 1975). Clearly, therefore, a breakdown in inter-cellular communication is likely to be based on molecular deterioration and need not be explained as a separate process (Timaras, 1975).

(3) *The basic, cross-linkage hypothesis is very general and states that an age-dependent accumulation of cross-linkages occurs between the side-groups of large polymers throughout the organism.* These linkages are brought about through the action of a wide variety of naturally occurring agents; for example, free radicals or any chemical having an odd number of electrons and aldehydes. Cross-linkage between the side-groups of the larger molecules modifies their shape and thus their biological effect. Cross-linkages may also bring about the association of small molecules, the attachment of small to large molecules and even the attachment of molecules to cellular organelles.

Much of the early work on cross-linking was carried out on extra-cellular materials like collagen and intracellular 'ageing pigment'. The extent to which collagen becomes cross-linked certainly increases with age (Verzar, 1963; Hall, 1976) but most of this appears to take place during development and maturation rather than during the senescent period of the life-cycle. Lipofuscin, the so-called 'age pigment', accumul-ates with age in the cells of many tissues (Strehler, 1962). This pigment, which is the product of cross-linking reactions, does not appear to damage cells except when it accumulates to such an extent that it interferes with the normal structure and functioning of the cell (Zeman, 1971). However, the occurrence of such obvious materials suggests that other, less obvious, cross-linking products may accumulate with age to the detriment of the tissue. There is, for example, a suggestion of progressive accumulation of cross-linking in the chromatin (von Hahn, 1970; Cutler, 1975; Tas, 1977), evidenced by: (1) a decrease in extractability of the proteins from DNA in aged cells; (2) an increase in the thermal stability of the chromatin with age; (3) an age-dependent attenuation of the template capacity and percentage transcription of the chromatin. The relationship of such changes to the ageing process has not been established, but some involvement is suggested by the fact that the longevity of certain laboratory animals has been extended by growing them on a diet rich in antioxidants which inhibit cross-linkage by autoxidation (Tappel, 1968; Tappel *et al.,* 1973; Kohn, 1976; Harman, 1972). Vitamin E, which has well-documented antioxidant properties, has been shown to increase the longevity of nematodes (Epstein and Gershon, 1972) and human tissue-culture cells (Packer and Smith, 1974). The biological role of vitamin E has not so far been established (Green, 1972) but it has been suggested that it may be a natural antioxidant (Hoffer and Roy, 1975). Another important study on antioxidants was that carried out by Comfort (1974) on mice. He was able to increase significantly the maximum life-span of these animals by adding 0.5% W/w ethoxyquinone, a powerful antioxidant, to the diet. On the negative side, however, it has to be noted that several studies on antioxidants have failed to evoke any effect at all (Cutler 1975).

There is, therefore, considerable potential for molecular damage both in cellular and intercellular materials. The evidence associating such damage with senescence is far from straightforward, but this is presumably what would be expected if ageing is based on a system of multiple causation of a stochastic kind. However, there are several major difficulties which a theory of this kind still has to face. Firstly, given that specific mechanisms seem to have evolved at both a molecular and cellular level to identify and remove or repair damage, how does damage accumulate with age? Secondly, how can the predictability of the ageing process be reconciled with a stochastic basis? Finally, how can some organisms apparently escape the accumulation of damage and hence ageing? What follows is an attempt to solve some of these problems, and is based largely on Calow (1978b).

(1) *On the accumulation of damage,* Orgel's 'Catastrophe Theory' tries to eliminate the problem by postulating a positive feedback system which brings about an acceleration in the build-up of malformed molecules (Fig. 8.6). This will only work, though, provided a certain level of errors is maintained despite repair. Orgel (1970) realized this and corrected his

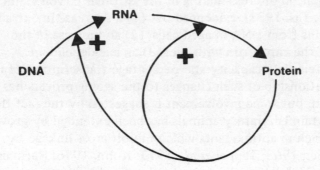

Fig. 8.6 Positive feedback in Orgel's Catastrophe Theory.

theory as follows: Let C_n = error frequency in the nth generation of the protein molecules produced by the $n-1$th generation of proteins in a cellular system; R = residual error frequency; α = proportionality constant between the errors in the synthetic apparatus and in the freshly synthesized proteins; then:

$$C_{n+1} = R + \alpha C_n \qquad\qquad 8.4$$

and if
$$C_0 = 0$$

$$C_n = R (1 + \alpha + \alpha^2 + \ldots\ldots + \alpha^{n-1}) \qquad\qquad 8.5$$

Only if $\alpha > 1$ will C_n increase indefinitely towards a catastrophe. If $\alpha \gg 1$, C_n will increase exponentially. Otherwise, as will be the case if there is repair, C_n will tend to a steady-state defined by $R/(1-\alpha)$ and there will be no catastrophe (see also Hoffman, 1974; Kirkwood and Holliday, 1975; Goel and Ycas, 1975).

The other alternative, of course, is that repair mechanisms themselves become impaired. A reduction in the fidelity of DNA repair by polymerases has already been noted. Turnover and replication processes, which are important in the replacement of damaged structures, may also slow down with age, thus allowing the accumulation of mistakes and damage. Some molecules, like nucleic acids and collagen, turn over at only a very slow rate under normal conditions (Section 4.2) and there is evidence for a general deceleration in molecular turnover with age. For example, nail and hair growth, which provide indices of the turnover of materials within the organism, slow down with age (e.g. Hamilton *et al.*, 1955). The general intensity of protein synthesis and breakdown is high in the newborn and declines rapidly with normal growth in man (Waterlow, 1967; Winterer *et al.*, 1976) and rodents (Waterlow and

Stephen, 1967; Yousef and Johnson, 1970; Johnson and Strehler, 1972; Soltesz *et al.,* 1973; Macieira-Coelho and Loria, 1974). However, the story is not a simple one since though whole body turnover reduces in these groups when measured in terms of body mass, it may actually rise when measured in terms of cell number or protein mass (Winterer *et al.,* 1976). Even so, it is now certain that protein turnover in specific body compartments, particularly in muscles, reduces with age (Young, 1976). Also, molecular turnover is usually well correlated with metabolic rate and on a per unit weight basis this invariably reduces with age and size (Kleiber, 1975).

There is an indisputable slowing down of overall cellular as well as molecular turnover during development, as differentiated cells are generally unable to divide (Bullough, 1967). Both *in vivo* and *in vitro* cells can apparently differentiate themselves to death (Martin *et al.,* 1975). At the same time there is a slowing down of mitosis due essentially to a prolongation of G_1 and G_2 (Lesher *et al.,* 1961 a and b; Thrasher, 1967; Lesher and Sacher, 1968; Gelfant and Smith, 1972; Ryan *et al.,* 1974, Gahan and Hurst, 1976).

(2) *There are now several possible ways of reconciling a predictable, ageing process with the accumulation of randomly-acquired, molecular damage.* Either the deterioration of repair mechanisms might be programmed by a positive act of selection in the way envisaged by Burnet (1974) or it may arise as a side-effect of traits selected for their positive contribution to fitness in the evolutionarily competent part of the life cycle, but which entail suboptimal repair as a side-effect. The first hypothesis has already been rejected on the basis of evolutionary logic so on *a priori* grounds the second hypothesis would appear to be most viable.

Kirkwood (1977) has formulated a very general hypothesis of the second kind. He suggests that repair is metabolically expensive (and this is certainly true of turnover; see Section 2.4) and that the full expense, needed for complete repair, cannot always be justified (in terms of selection and fitness) in organisms that stand a good chance of dying anyway through accident, disease or predation – as most organisms do in the 'wild' (Section 8.1). In other words the organism should use all possible energy and resources (consonant with its survival to optimum reproductive value – see Section 7.3) to promote growth (Sections 2.7 and 5.1), survival during starvation or suboptimal feeding levels (Section 2.7) and reproduction (Sections 7.4 and 7.5), and ageing comes as a side-effect of this strategy. Kirkwood postulates that expensive, high fidelity repair must be characteristic of germ cells, since these cannot be allowed to age, and that the switch to submaximal repair comes at the time of the differentiation of the soma from the germ line — this would perhaps explain why division is stopped in differentiated cells (Section 3.2). However, there is no direct evidence, as yet, of more expensive repair processes in germ cells, and other mechanisms could be involved here to keep the germ line error-free (see Section 9.5). There is no reason, therefore, why the switch should not occur at other times in the life cycle; e.g. at the onset of reproduction, where resources are often channelled differentially into gametic production (Sections 7.3 and 7.5). Another possibility is that a 'rheostat' is involved rather than a 'switch' and that this diverts more energy from the soma to the reproductive processes with time (Calow, 1978b). Such a process is non-circular if the 'rheostat' is geared to developmental events, e.g. the decreasing growth rate, rather than ageing events. This would therefore explain the programme-like fashion by which turnover slows down with age, and would enable us to reconcile the stochastic occurence of errors with their predictable accumulation with time.

In summary, then, I suggest that repair, replacement and turnover are needed to keep the organismic system clear of damage and that when they occur at a level which is insufficient to do this the tissues age.

(3) *A corollary of these arguments is that systems in which turnover and selection are possible all the time ought to be potentially immortal.* There is a definite negative correlation between the degree of differentiation of a system and the occurrence of ageing, such that ageing becomes a more obvious feature of higher organisms (Table 16). Similarly, plants which have continuously active growing points are often apparently immortal (D'Amato, 1963; Woolhouse, 1972). In the higher plants there is also the possibility of selection, since cell divisions within the apical meristems give rise to daughter cells with different fates; some of the cells remain meristematic, whilst others ultimately die as they differentiate into vascular tissues and fibres or remain alive for a time as leaf or pith parenchyma. Ultimately, the leaves age and fall from the tree and the pith breaks down. It is not inconceivable, therefore, that either 'toxic materials' or the products of molecular damage are differentially transferred to the cells, which ultimately die, thus retaining a non-ageing, stem cell system (Sheldrake, 1974).

8.6 Conclusions

From the above it would seem that there are two elements to the ageing process; damage and repair. At a molecular level, damage of one kind or another is likely to occur throughout the life cycle but is removed by repair mechanisms. Repair implies replacement and replacement neccessitates turnover. Hence, molecular and cellular turnover are fundamental processes of repair, and I propose that senescence occurs not so much because of an increase in the frequency of damage with age, but because of a deterioration in the repair processes, particularly turnover. It is conceivable that in certain circumstances, where kin selection is proven, that such a deterioration could be programmed directly; but it would seem more plausible that it occurs as a side-effect of the selection of traits which promote fitness by switching material and energy from repair processes to other aspects of metabolism concerned directly with growth and reproduction.

9 The cycle reversed

C.M. Child once wrote a book with the provocative title: *Senescence and Rejuvenescence* (Child, 1915). It is very appropriate, however, to consider these two processes together as Child did, for any system which ages must become rejuvenated, in some sense of the word, if it is to persist. Both sexual and asexual reproduction, for example, must involve a kind of rejuvenation if their products are not to continue ageing from where the parents left off. Furthermore, in some organisms it is possible to reverse the developmental programme, and perhaps also to rejuvenate, artificially. Rejuvenation by reproduction and these contrived methods will be our main concern in this chapter.

9.1 Degrowth in triclads

In common with coelenterate medusae (Hadzi, 1910), nemertines (Dawydoff, 1910) and some ascidians (Huxley, 1926) most triclads can withstand long periods of starvation and during this time may shrink from an adult size back to, and sometimes beyond, their initial size at hatching (Fig. 9.1). Since shrunken worms resemble juveniles in both appearance and physiology (Child, 1915) it has been suggested that starvation not only takes worms back in size but that it also takes them back in age; in other words degrowth is supposed to reverse life's cycle.

Fig. 9.2 maps the quantitative changes in the body size of triclads which take place during growth, shrinkage (degrowth) and regrowth. Triclad size is expressed as plan area because this parameter can be measured with ease and accuracy and because it is well correlated with the weight of the worms. From the data it is possible to see that hatchlings grow in a typical sigmoid fashion, that the rate of degrowth falls in an exponential fashion with starvation time, and that regrowth tracks a continuously

124

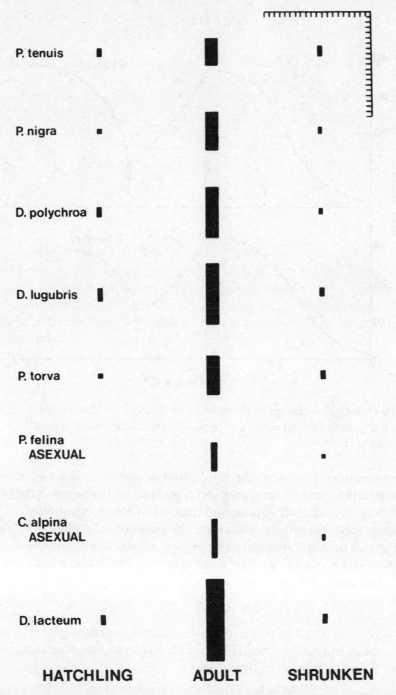

P. tenuis

P. nigra

D. polychroa

D. lugubris

P. torva

P. felina
ASEXUAL

C. alpina
ASEXUAL

D. lacteum

HATCHLING   **ADULT**   **SHRUNKEN**

Fig. 9.1 Relative size changes from birth to adulthood in triclads that are fed to satiation, and shrinkage from adult size to the minimum size before death in triclads that are completely starved. The scale is in millimeters.

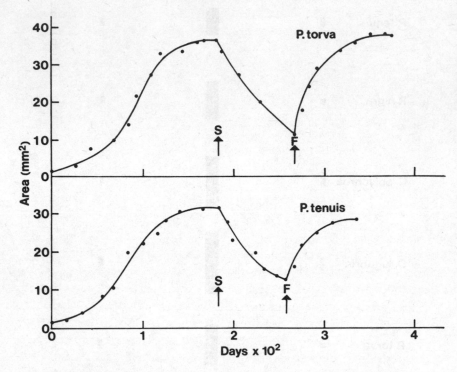

Fig. 9.2 Quantitative changes in the size of two species of triclad during growth, degrowth and regrowth. S = starve, F = feed. Size is measured as area — see text.

decelerating curve. It is clear, therefore, that degrowth is not a reversal of growth in the sense of following out a reversal of the sigmoid curve. Nevertheless, it might still represent a reversal of those metabolic processes involved in growth. Consider, for example, the simple metabolic growth model of von Bertalanffy (see review 1960) where growth in weight $(\mathrm{d}W/\mathrm{d}t)$ is modelled as the outcome of the difference between anabolic (B) and catabolic (C) processes:

$$\mathrm{d}W/\mathrm{d}t = B - C \qquad\qquad 9.1$$

B is usually considered to be a surface-dependent function $(fW^{0.67})$ and C a weight-dependent function (FW^1). Substituting and integrating equation 9.1 with respect to time gives:

$$W_t = W_\infty (1-\mathrm{e}^{-k\ (t-t_o)})^3 \qquad\qquad 9.2$$

where W_t = weight at time t; W_∞ = final size; t_o = theoretical point in time when $W_t = 0$; e = base of natural logarithms. Assuming that weight is functionally related to area raised to the power of 1.5, then equation 9.2 becomes

$$A_t = A_\infty (1-e^{-(k/3)(t-t_o)^2} \qquad 9.3$$

where the terms are as specified in equation 9.2 but are expressed in area units. Now during starvation, B drops out of equation 9.1 to give:

$$-dW/dt = FW^1$$

and
$$-dA/dt = \gamma A^1 \qquad 9.4$$

where $\gamma = F^{0.67}$. Integrating equation 9.2 with respect to time gives:

$$A_t = A_o e^{-\gamma t} \qquad 9.5$$

Equations 9.2 and 9.3 model rising sigmoid curves, but equation 9.5 models a response that falls at an exponential rate with time. Removing the anabolic term from von Bertalanffy's model therefore transforms a sigmoid growth curve to an exponential degrowth curve. Hence, metabolically, degrowth can be considered as a process in which growth is put in reverse. The more limited size reductions which accompany starvation in higher organisms are also of an exponential form (Morgulis, 1923) though a more complex response may occur in man (Forbes, 1970) and senile shrinkage may even track out a sigmoid trajectory (Needham, 1962).

It is, of course, still an open question as to whether the growth reversal in triclads leads to an age reversal. One of the problems is that we know very little about the morphological correlates of ageing in triclads and from this point of view find it difficult to judge, simply from visual evidence, whether rejuvenation has occurred. Classically, the problem has been investigated in terms of the influence that degrowth has on metabolism and I consider this evidence in the next section. Clearly, the strongest basis for suggesting that degrowth effects rejuvenation would come from an investigation of the effect it has on worm life-span and this I consider in Section 9.3.

9.2 Physiological evidence for rejuvenation in triclads

Metabolism per unit weight invariably reduces as organisms get older and larger (Section 2.1). On the assumption that an organism's resistance to dilute cyanide is a good index of its metabolic rate, Child (1915) measured this property in growing and degrowing worms. He found that

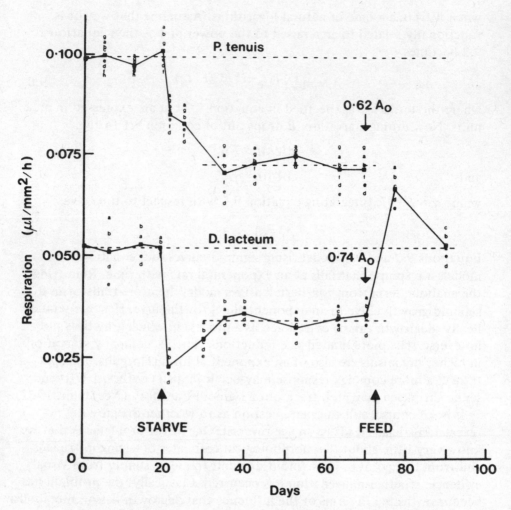

Fig. 9.3 Changes in the respiration of two tricald species during starvation.
A_0 = initial size before starvation. (With permission from Calow and Woollhead, 1977b).

susceptibility to cyanide decreased as the animal grew but increased as it degrew. Similarly, Hyman (1919, 1920) found a reduction in oxygen consumption per unit weight of tissue as triclads grew older and larger but a reduction in weight-specific oxygen consumption during degrowth. The results of both Child and Hyman suggest, therefore, that triclads become physiologically rejuvenated during starvation. However, other workers (Allen, 1919 and Pederson 1956) have recorded either no change in the oxygen uptake per unit weight of triclad tissue or a fall in uptake.

Recently with A.S. Woollhead (Calow and Woollhead, 1977) I have made the following observations which bear on the problem. Firstly, the relationship between plan area and dry weight in two triclads, *Polycelis tenuis* and *Dendrocoelum lacteum*, is allometric, with area (A) being related to weight (W) by $W \propto A^{1.5}$. Secondly, the relationship between worm plan area and respiration (R) is isometric: $R \propto A^1$. Hence respiration per unit area (R/A) should be constant and independent of size. Thirdly, during starvation R/A falls from a steady-state but ultimately stabilizes around a new steady-state (Fig. 9.3). Therefore, since after the initial adjustment, R/A remains constant but A/W increases with shrinkage, then R/W, the respiratory rate per unit weight, will also increase with shrinkage. This conclusion supports the view of Child and Hyman and it can, therefore, be argued that degrowth rejuvenates metabolism. However, in that it is possible to explain most of these changes solely on the basis of alterations in the size of worms, particularly in the area per weight (A/W) ratio, they need not support the idea of a true rejuvenation of the system in the sense of an extension of life. In the absence of good criteria for distinguishing between 'old' and 'young' features of triclad tissue it is therefore necessary to refer to actuarial data for an answer to this problem.

9.3 Actuarial evidence for rejuvenation

One problem with an actuarial approach to the study of the ageing and rejuvenation of triclads is that these animals may have long laboratory lives of up to twenty years (Balázs and Burg 1962; Haranghy and Balázs, 1964—66). Indeed, because of this it has been suggested that triclads do not age in the normal way (Comfort, 1964). Other investigators, however, have found a definite age-associated deterioration in several of the metabolic and morphological features of laboratory raised worms and this is a clear indication of senescence (Balázs and Burg, 1962; Haranghy and Balázs, 1964—66). In a series of experiments I have tried to overcome the difficulty presented by long life by exploiting the well-known life-shortening effect of high energy, γ, irradiation. (Calow, 1977c).

Fig. 9.4 illustrates the average survival time of different aged triclads after irradiation. In general, young (3 months old) and old (24 months old) triclads were more susceptible to irradiation than middle-aged (12 months) triclads. Hence, triclads became less susceptible to irradiation as they grew larger and older, but once they reached maturity and ceased to grow this trend was reversed. However, the relationship between

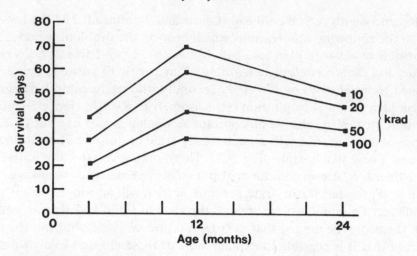

Fig. 9.4 Post-irradiation life-span in *D. lacteum* of various ages. (With permission from Calow, 1977c).

survival after irradiation and age in the older worms was dose-dependent. That is to say, the difference between the survival of middle- and old-aged worms was most marked at 10 krad and least marked at 100 krad. These results can be explained if it is assumed that irradiation has two effects; one which contributes to (i.e. accelerates) the normal processes of ageing and the other which produces non-specific life-shortening (i.e. injures). It is well known that young, growing organisms, with actively dividing tissues, are more susceptible to injury from irradiation than their non-growing counterparts (Alexander, 1957; Strehler, 1959). Hence, the reduction in the life-shortening effect of the γ rays as the triclads grew from 'youth' to 'middle-age' can be ascribed to reductions in the injurious effects of the irradiation. Alternatively, the life-shortening effect of irradiation, between middle- and old-age, when the turnover of tissues is much reduced, can be ascribed to a predominant ageing effect. A similar interpretation can be applied to the results of Hursh and Casarett (1956) and Blair (1959) on mammals, but the data of Kohn and Kallman (1956) on *Drosophila* are more equivocal.

As dose increases, it is likely that both the ageing and injurious effects of irradiation increase but that the limits of the ageing effect are reached before the limits of the injurious effect. Since the injurious effect is likely to be independent of age in non-growing worms, then the age-independent response above 20 krad can be ascribed to injury. Therefore, the ageing response is only manifest at 20 krad and below.

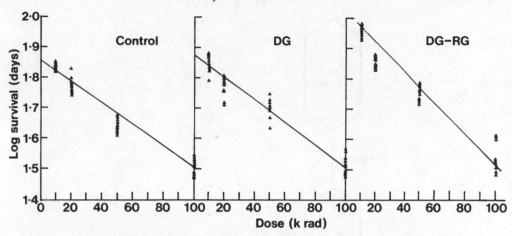

Fig. 9.5 Relationship between post-irradiation life-span and dose levels in untreated worms (Control), worms starved to 50% initial, adult size (DG) and worms regrown to full size (DG−RG). The slope for the DG−RG line is about 1.3 times greater than that for the others and this is significant. (With permission from Calow, 1977c).

Dose-response curves, plotting the logarithm of survival against level of irradiation, are presented in Fig. 9.5 for full-size, degrown (1/2 full-size) and regrown, adult *D. lacteum.* These relationships are linear, but the slope of the line for the regrown worms differed significantly from that of the others. This was because the life-spans of worms treated with more than 20 krad were all similar irrespective of the previous feeding regime, whereas the life-spans of regrown worms treated with doses of 20 krad and less were significantly longer than the life-spans of the other worms. It is feasible, therefore, that degrowth *plus* regrowth had a rejuvenating effect and thereby retarded the ageing effect of irradiation. Rejuvenation would be unlikely to influence the injurious effect of irradiation so no extension of post-irradiation life-span would be expected at dose levels above 20 krads. On its own, degrowth seemed to have no effect whatsoever. Hence, if rejuvenation occured at all the active agent must have been regrowth.

Fig. 9.6 is a pictorial model which tries to explain this ageing and rejuvenation in triclads. Ageing occurs through the accumulation of some factor that I call *a*, which may either be non-specific molecular damage as envisaged in Section 8.5 or perhaps even a toxin. Rejuvenation involves a reduction in the density of *a* and irradiation accelerates its accumulation. Reductions in the density of *a* might occur if *a*-laden cells and molecules were preferentially catabolized during degrowth or if *a*-laden tissues were diluted out or flushed out during regrowth. The experimental

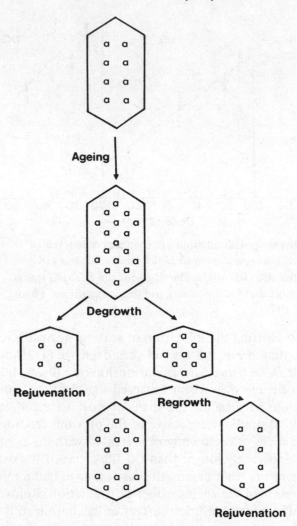

Fig. 9.6 Model of rejuvenation in triclads. See text for further explanation.
(With permission from Calow, 1977c).

results support the latter rather than the former possibility. The biological
basis of this conclusion may be as follows: Triclads contain a stock of
totipotent stem cells, the neoblasts, which are latent for most of the time,
and therefore not susceptible to damage, but which are called into play
to replace lost tissue after wounding or asexual fission. There is also
evidence for a dramatic increase in the mitotic activity of these cells after
the refeeding of starved worms (Baguña, 1974). Hence, during regrowth
the neoblasts may produce metabolically 'young' cells, which dilute out
the pre-existing 'old' ones.

In a series of publications Lange (1966 and 1967) has also suggested that ageing in triclads is associated with a reduction in the density of neoblasts in the tissue and in consequence with reduced tissue turnover. This fits in with the model depicted in Fig. 9.6 except that instead of putting the emphasis on turnover *per se* as an element of senescence and rejuvenation I focus attention on the effects of reduced turnover in senescence (an increase in 'a') and increased turnover in rejuvenation (a reduction in 'a'). Clearly, a complete explanation of senescence and rejuvenation must take into account both the dynamics of tissue turnover in the organism and its effects.

9.4 Rejuvenation by asexual fission

The hypotheses put forward in the last section are also borne out by 'natural experiments'. Some species of triclads and rhabdocoels reproduce by binary fission. There is no degrowth here, only regrowth of the fragments after fission. However, there must be rejuvenation otherwise species depending exclusively on asexual fission as a means of reproduction would ultimately die out. Hence regrowth after fission rejuvenates, presumably in a similar way to regrowth after shrinkage, and this could explain along the 'neoblast system' evolved in the first place.

Each fission product need not be rejuvenated to an equal extent. For example, in a series of classical experiments on the asexual rhabdocoel *Stenostomum incaudatum*, Sonneborn (1930) found that lineages derived from successive heads soon ceased to divide and died out. Alternatively, lineages derived from tails continued to divide and showed apparent immortality. These differences may be a function of the extent to which remaining tissues turn over and are replaced after fission. Tails become completely reorganized and yet heads retain the old brain and specialized head apparatus. Also interesting from this point of view are the results of Haemerling (1924) on the oligochaete *Aeolosoma*. He found that only the specialized anterior of these worms aged — new worms being produced from the posterior end. Also Harms (1949) was able to rejuvenate old serpulids by removing old heads and grafting on young ones. Hence, as suggested in Section 8.5, senescence seems to be associated with specialization and with a lack of tissue turnover.

9.5 Rejuvenation by sexual reproduction

Reproduction of any kind must, by definition, rejuvenate. The way this occurs in sexual reproduction may be similar, in broad principles, to the way it occurs after transection and fission. In sexual reproduction, though, development procedes from a zygote — the totipotent stem cell formed from the fusion of gametes. These sex cells, like the neoblasts, are kept in an inactive state until required. Furthermore, gonadal tissue becomes separated off at an early stage in development and remains latent for most of the time before maturity. Since metabolic activity is known to have a damaging effect on cells and molecules these strategies presumably protect the genetic information carried by the gametes. Other devices associated with sexual reproduction might conceivably rejuvenate. First, it is possible that the genetic assortment associated with meiosis might be involved. Certainly, senescent cultures of asexual micro-organisms can be rejuvenated by the onset of sexuality. Secondly, asymmetrical divisions of the egg mother cell often lead to the formation of a number of cells which, though supportive in early development, ultimately die. Hence, 'bad' cytoplasm might be left behind in these short-lived cells (Sheldrake, 1974). There is no.comparable loss in male gametogenesis, but this is to be expected since almost all the cytoplasm from which the embryo develops is derived from the egg and not the sperm. Third, natural selection itself must play a part in ensuring that only the fit develop and go on to reproduce. Contrary to the assumption that, given suitable conditions, the majority of gametes would form zygotes that would develop normally (Saunders, 1970) there is now strong evidence that many imperfect gametes are formed through copying errors and that these are filtered out by rigorous selection on the gamete, the egg and the early developmental stages (Cohen, 1977). Fourthly, and finally, there may, as Kirkwood (1977) suggested (Section 8.5), be more accurate repair mechanisms in germinal cells — though there is no direct evidence for this as yet.

From all this data on rejuvenating mechanisms one simple conclusion forces itself upon us; that rejuvenation involves cellular turnover and selection. Furthermore, those cells that remain juvenile without turning over, the neoblasts under normal conditions, and the gametes before maturity, probably do so as latent, non-functional cells with low metabolic rates. These conclusions reinforce the interpretation of senescence given in the last chapter — that the ageing process occurs through the absence of cellular turnover in specialized, functionally competent, metabolizing tissues. Furthermore, these conclusions have definite implications for our interpretation and understanding of life's cycle which we shall pursue in the next, final section.

Part five Summary and conclusions

Living systems are most distinguishable from non-living ones in the way
that they avoid the consequences of the second law of thermodynamics.
The basis for their persistence, despite the entropic forces of the world,
is the separation of a protected and thermodynamically stable programme
of instructions (the genome) from an active, meta-stable, working-part
(the phenotype). Damaged, phenotypic components can thus be replaced
according to these genetic instructions, and when this fails the organismic
system as a whole can be reproduced from replicated copies of the genome.
The success of such a process depends upon the key property of non-
reciprocity (Calow, 1976) which is manifest as the unidirectional flow of
information from the genome to the phenotype and the control of the
expression of the genome in each cell through a switching system, not
through the input of new and detailed information. Without such a property it would be feasible for maladaptive changes, which arise in the
phenotype as a result of its metabolic and physical activity, to become
written into the genome. On the other hand, non-reciprocity puts certain
constraints on the mechanism of evolution and on the control of cellular
metabolism; Lamarckism is precluded and cellular control depends ultimately on a series of binary decisions — switch off or switch on. It might
be thought that such constraints put limitations on the complexity and
diversity that can be achieved by living systems, but even simple nets
of binary switches can show surprising complexity and stability
(Kauffman, 1969).

The need for a continuous process of repair within the active phenotype leads to one of the most fundamental features of organisms; that
of turnover. Repair necessitates replacement, and replacement leads to
measurable fluxes in the tissues, cells and molecules of the organism.
Organisms are part of a cycle, their life cycle, but they are also made up

135

of a series of cycles within cycles. Degraded macro-molecules are continuously replaced with freshly synthesized molecules and both the subcellular organelles and the cells themselves turn over in much the same way. In the early parts of the life cycle such processes of regeneration are able to keep pace with the degenerative processes within the system but in later life this may change. The progressive accumulation of damage at one level may lead to failure in the systems in the level above until the organism as a whole ages and dies. Persistence, then, is dependent on organismic reproduction, the 'ratchet of life', which prevents particular living systems and life in general from slipping back into disorder.

If they do not die from disease, predation or accident most organisms die from 'old age'. In the end, therefore, the success of a particular kind of organismic system depends on how good it is at reproducing itself – that is, at transmitting replicas of the instructions it carries to future generations. At the same time, how successful it is at reproducing itself depends upon its own survival. In other words, the genetic fitness of a system, measured in terms of progeny production, depends on its own, individual fitness, measured in terms of ability to find food and avoid accident, disease and predation. In the final analysis both fitnesses depend, to a large extent, on how good the organism is at partitioning the energy and resources available to it amongst the several demands of maintenance, growth and reproduction. It is my firm belief that understanding how organisms optimize the utilization of their energy and resource input in order to maximize their output of genetic information comes close to *the* fundamental goal of organismic biology. This idea has been my main concern throughout the book and in a more rigorous form, for example incorporating optimality principles (Rosen, 1967), will undoubtedly give direction to future studies of metabolism and development.

References

Abelson, P.H. (1957). Some aspects of palaeobiochemistry. *Ann. N.Y. Acad. Sci.,* **69**, 276–285.

Abrahamson, W.G. and Gadgil, M. (1973). Growth, form and reproductive effort in Goldenrods (Solidago, Compositae). *Am. Nat.,* **107**, 651–661.

Adelman, R.C. (1970). An age-dependent modification of enzyme regulation. *J. biol. Chem.,* **245**, 1032–1035.

Adelman, R.C. (1975). Impaired homonal regulation of enzyme activity during ageing. *Fedn. Proc. Fedn. Am. Socs. exp. Biol.,* **34**, 179–185.

Adelman, R.C., Freeman, C. and Cohen, B.S. (1972). Enzyme adaptation as a biochemical index of development and ageing. *Adv. Enzymol.,* **10**, 365–382.

Alexander, P. (1957). Accelerated aging–a long term effect of exposure to ionizing radiations. *Gerontologia,* **1**, 174–193.

Alexander, R.McN. (1967). *Functional Design in Fishes.* Hutchinson University Library, London.

Allen, G.D. (1919). Quantitative studies on the rate of respiratory metabolism in planaria. II The rate of oxygen consumption during starvation, feeding, growth and regeneration in relation to the method of susceptibility to potassium cyanide as a measure of rate of metabolism. *Am. J. Physiol.,* **49**, 420–475.

Altenberg, E. (1934). A theory of hermaphroditism. *Am. Nat.,* **68**, 88–91.

Anderson, J.F. (1970). Metabolic rates of spiders. *Comp. Biochem. Physiol.,* **33**, 51–72.

Atkinson. D.E. (1972). The adenylate energy charge in metabolic regulation. In: *Horizons of Bioenergetics,* ed. San Pietro, A. and Gest, H. Academic Press, New York and London, pp. 83–96.

Baguña, J. (1974). Dramatic mitotic response in planarians after feeding and a hypothesis for the control mechanism. *J. exp. Zool.,* **190**, 117–122.

Baird, M.B. and Massie, H.R. (1975). A further note on the Orgel Hypothesis and senescence. *Gerontology,* **21**, 240–243.

Baird, M.B., Samis, H.V., Massie, H.R. and Zimmerman, J.A. (1975). A brief argument in opposition to the Orgel Hypothesis. *Gerontology*, **21**, 57–63.

Balázs, A. and Burg, M. (1962). Span of life and senescence of *Dugesia lugubris. Gerontologia*, **6**, 227–236.

Barnes, H. and Crisp, D.J. (1956). Evidence of self-fertilization in certain species of barnacles. *J. mar. biol. Ass. U.K.,* **35**, 631–639.

Barton, R.W. and Young, M.K. (1975). Low molecular weight DNA polymerase; decreased activity in spleens of old BAB/c mice. *Mech. Ageing and Dev.,* **4**, 123–136.

Baserga, R.C. (1976). *Multiplication and Division in Mammalian Cells.* Marcel Dekker, New York and Basel.

Beauchene, R., Roeder, L. and Barrows, C. (1967). The effect of age and of ethionine feeding on the ribonucleic acid and protein synthesis of rats. *J. Geront.,* **22**, 318–324.

Bertalanffy, L. von. (1960). Principles and theory of growth. In: *Fundamental Aspects of Normal and Malignant Growth*, ed. Nowinski, W.W. Elsevier, Amsterdam, London, New York, Princeton, pp. 137–259.

Bertalanffy, L. von and Pirozynski, W.J. (1951). Tissue respiration and body size. *Science,* **113**, 599–660.

Blair, H.A. (1959). Data pertaining to shortening of life-span by ionizing radiation. *US Atomic Energy Commission Report.* **UR–442**.

Blower, G. (1969). Age-structures of millipede populations in relation to activity and dispersion. In: *The Soil Ecosystem* ed. Sheal, J.G. Syst. Ass. Publ. No. 8, pp. 209–216.

Blum, M.S. (1955). *Time's Arrow and Evolution.* Harper, New York.

Boardman, N.K. (1977). Comparative photosynthesis of sun and shade plants. *A. Rev. Pl. Physiol.,* **28**, 355–377.

Bolla, R. and Brot, N. (1975). Age-dependent changes in enzymes involved in macromolecular synthesis in *Turbatrix aceti. Archs. Biochem. Biophys.,* **169**, 227–236.

Bonner, J.T. (1965). *Size and Cycle.* Princeton University Press, Princeton N.J.

Borsook, H. (1950). Protein turnover and incorporation of labelled amino acids. *Physiol. Rev.,* **30**, 206–219.

Bozouk, A.N. (1976). Testing the protein error hypothesis of aging in *Drosophila. Expl. Geront.,* **11**, 103–112.

Brett, J. (1971). Satiation time, appetite and maximum food intake of sockeye salmon (*Oncorhynchus nerka*). *J. Fish. Res. Bd Can.,* **28**, 409–415.

Broda, E. (1975). *The Evolution of the Bioenergetic Processes.* Pergamon Press, Oxford, New York, Toronto, Sydney, Paris.

Brodie, P.F. (1975). Cetacean energetics, an overview of intraspecific size variation. *Ecology,* **56**, 152–161.

Bucher, N., Swaffield, M., Moolten, F. and Schrock, T. (1969). Early events in hepatic regeneration. In: *Biochemistry of Cell Division*, ed. Baserga, R.C. Thomas, Springfield, Illinois, pp. 139–154.

Bullough, W.S. (1952). The energy relations of mitotic activity. *Biol. Rev.,* **27,** 133–168.

Bullough, W.S. (1967). *The Evolution of Differentiation.* Academic Press, London and New York.

Bullough, W.S. (1969). Epithelial repair. In: *Repair and Regeneration,* ed. Dunohy, J.E. and van Winkle, W. McGraw-Hill, New York.

Bullough, W.S. and Laurence, E.B. (1960). The control of epidermal mitotic activity in the mouse. *Proc. R. Soc.,* **151B,** 517–536.

Burlew, J.S. (1964). *Algal Culture.* Carnegie Institute of Washington **600,** Washington D.C.

Burnet, M. (1974). *Intrinsic Mutagenesis: A Genetic Approach.* M.T.P., Lancaster.

Cairns-Smith, A.G. (1971). *The Life Puzzle.* Oliver and Boyd, Edinburgh.

Calow, P. (1973). On the regulatory nature of individual growth: some observations from freshwater snails. *J. Zool. Londs.,* **170,** 415–428.

Calow, P. (1974). Some observations on locomotory strategies and their metabolic effects in two species of freshwater gastropods, *Ancylus fluviatilis* Müll. and *Planorbis contortus* Linn. *Oecologia,* **16,** 149–161.

Calow, P. (1975). The respiratory strategies of two species of freshwater gastropods (*Ancylus fluviatilis* Müll. and *Planorbis contortus* Linn.) in relation to temperature, oxygen concentration, body size and season. *Physiol. Zool.,* **48,** 114–129.

Calow, P. (1976), *Biological Machines.* Arnold, London.

Calow, P. (1977a). Ecology evolution and energetics. *Adv. Ecol. Res.,* **10,** 1–62.

Calow, P. (1977b). Conversion efficiencies in heterotrophic organisms. *Biol. Rev.,* **52,** 385–409.

Calow, P. (1977c). Irradiation studies on rejuvenation in triclads. *Expl Geront.,* **12,** 173–179.

Calow, P. (1978a). The evolution of life cycle strategies in freshwater gastropods. *Malacologia,* in press.

Calow, P. (1978b). Bidders hypothesis revisited: a solution to some key problems associated with the general molecular theory of ageing. *Gerontology,* in press.

Calow, P. and Jennings, J.B. (1974). Calorific values in the phylum Platyhelminthes: the relationship between potential energy, mode of life and the evolution of entoparasitism. *Biol. Bull.,* **147,** 81–94.

Calow, P. and Jennings, J.B. (1977). Optimal strategies for the metabolism of reserve materials in microbes and metazoa. *J. theor. Biol.,* **65,** 601–603.

Calow, P. and Woollhead, A.S. (1977a). Locomotory strategies in freshwater triclads and their effects on the energetics of degrowth. *Oecologia,* **27,** 353–362.

Calow, P. and Woollhead, A.S. (1977b). The relationship between ration reproductive effort and age-specific mortality in the evolution of life-history strategies— some observations on freshwater triclads. *J. Anim. Ecol.,* **46,** 765–781.

Charnov, E.L. and Krebs, J.R. (1973). On clutch-size and fitness. *Ibis,* **116,** 217–219.

Charnov, E.L. and Schaffer, W.M. (1973). Life history consequences of natural selection: Cole's result revisited. *Am. Nat.,* **107,** 791–793.

Child, C.M. (1951). *Senescence and Rejuvenescence.* University of Chicago Press, Illinois.

Clarke, C.A. and Sheppard, P.M. (1966). A local survey of the distribution of industrial melanic forms in the moth *Biston betularia* and estimates of the selective values of these in an industrial area. *Proc. R. Soc.,* **165B**, 424–439.

Cody, M. (1966). A general theory of clutch size. *Evolution,* **20**, 174–184.

Cohen, J. (1977). *Reproduction.* Butterworths, London.

Cole, L.C. (1954). The population consequences of life-history phenomena. *Q. Rev. Biol.,* **29**, 103–137.

Comfort, A. (1956). *The Biology of Senescence.* Routledge and Kegan Paul, London.

Comfort, A. (1964). *Ageing; the Biology of Senescence.* Routledge and Kegan Paul, London.

Comfort, A. (1974). The position of ageing studies. *Mech. Ageing and Dev.,* **3**, 1–31.

Conaway, C.H. (1971). Ecological adaptation and mammalian reproduction. *Biology of Reproduction,* **4**, 239–247.

Cook, L.M. (1971). *Coefficients of Natural Selection.* Hutchinson, London.

Crow, J.F. and Kimura, M. (1965). Evolution in sexual and asexual populations. *Am. Nat.,* **99**, 439–450.

Crow, J.F. and Kimura, M. (1971). *Population Genetics.* Princeton University Press, Princeton. N.J.

Cummins, K.W. and Wuychek, J.C. (1971). Caloric equivalents for investigations in ecological energetics. *Communs. Int. Assoc. Theoret. Appl. Limnol.,* **18**, 1–158.

Curtis, H.J. (1971). Genetic factors in ageing. *Adv. Genet.,* **16**, 305–324.

Curtis, H.J. and Tilley, J. (1971). The life-span of dividing mammalian cells *in vivo. J. Geront.,* **26**, 1–7.

Cutler, R.G. (1975). Cross-linkage hypothesis of ageing: DNA adducts in chromatin as a primary ageing source. In: *Cell Impairment, Ageing and Development,* eds. Cristofalo, V.I. and Holěckova, E. Plenum Press, New York.

D'Amato, F. (1963). Cytological and genetic aspects of ageing. In: *Genetics Today,* **2**, *Proc. XI. Int. Congr. Genetics,* ed. Geerts, S.J. Pergamon Press, London.

Danot, M., Gershon, H. and Gershon, D. (1975). The lack of altered enzyme molecules in 'senescent' mouse embryo fibroblasts in culture. *Mech. Ageing and Dev.,* **4**, 289–299.

Dawkins, R. (1976). *The Selfish Gene.* Oxford University Press, Oxford.

Dawydoff, C. (1910). Restitution von Kopfstücken, die vor der Mundöffnung abgeschnitten waren, bei den Nemertinen (*Lineus lacteus*). *Zool. Anz.,* **36**, 1–6.

Doctors van Leeuwen, W.M. (1931). Vogelbesuch an den Blüten von einigen *Erythrina* – Arten auf Java. *Annls. Jard. bot. Buitenz.,* **42**, 47–96.

Dugdale, A.E. and Payne, P.R. (1977). Pattern of lean and fat deposition in adults. *Nature, Lond.,* **266**, 349–351.

Ebling, F.J. (1955). Endocrine factors affecting cell replacement and cell loss in epidermis and sebaceous glands in the female albino rat. *J. Endocr.,* **12**, 38–49.

Edney, E.B. and Gill, R.W. (1968). Evolution of senescence and specific longevity. *Nature*, Lond., **220**, 881–882.

Eisenbach, E.S., Shimron, F. and Yagil, G. (1976). The effect of age on the regulation of glucose-6-phosphate dehydrogenase in the liver. *Expl. Geront.*, **11**, 63–72.

Elliott, J.M. (1975). The growth rate of brown trout (*Salmo trutta L.*) fed on reduced rations. *J. Anim. Ecol.*, **44**, 823–842.

Emlen, J.M. (1973). *Ecology: An Evolutionary Approach*. Addison-Wesley, Reading (Mass.) and London.

Enesco, M. and Leblond, C.P. (1962). Increase in cell number as a factor in the growth of organs and tissues of the young male rat. *J. Embryol. exp. Morph.*, **10**, 550–562.

Epstein, J. and Gershon, D. (1972). Studies on ageing in nematodes IV. The effect of antioxidants on cellular damage and life-span. *Mech. Ageing and Dev.*, **1**, 257–264.

Epstein, J., Williams, J.R. and Little, J.B. (1974). Rate of DNA repair in progeric and normal human fibroblasts. *Biochem. biophys. Res. Commun.*, **59**, 850–857.

Eshel, I. and Feldman, M.W. (1970). On the evolutionary effect of recombination. *Theoret. Pop. Biol.*, **1**, 88–100.

Everett, J.W. (1961). The mammalian female reproductive cycle and its controlling mechanisms. In: *Sex and Internal Secretions* ed. Young, W.C., 3rd Ed., Vol. II Williams and Wilkins, Baltimore, pp. 497–555.

Faegri, K. and van der Pijl, L. (1966). *The Principles of Pollination Ecology*. Pergamon Press, New York.

Ficq, A. and Pavan, C. (1957). Autoradiography of polytene chromosomes of *Rhynchosciara angelae* at different stages of larval development. *Nature*, Lond., **180**, 983–984.

Fisher, R.A. (1930). *The Genetical Theory of Natural Selection*. Oxford University Press, Oxford.

Forbes, G.B. (1970). Weight loss during fasting; implications for the obese. *Am.J. clin. Nutr.*, **23**, 1212–1219.

Ford, P.J. (1976). Control of gene expression during differentiation and development. In: *The Developmental Biology of Plants and Animals*, eds. Graham, C.F. and Wareing, P.F. Blackwells, Oxford, pp. 302–345.

Frey-Wyssling, A. (1953). *Submicroscopic Morphology of Protoplasm*. Elsevier, Houston.

Fukuda, M. and Sibatani, A. (1953). Biochemical studies on the number and the composition of liver cells in postnatal growth of the rat. *J. Biochem.*, (Tokyo), **40**, 95–110.

Gadgil, M. and Bossert, W.H. (1970). Life historical consequences of natural selection. *Am. Nat.*, **104**, 1–24.

Gadgil, M.D. and Bossert, W.H. (1970). Life historical consequences of natural evidence from wild flowers and some theoretical considerations. *Am. Nat.*, **106**, 14–31.

Gahan, P.B. and Hurst, P.R. (1976). Effect of ageing on the cell cycle of *Zea mays*. *Ann. Bot.*, **40**, 887–890.

Garrow, J.S. (1974). *Energy Balance and Obesity in Man*. North-Holland, Amsterdam, London.

Gee, J.M. and Williams, G.B. (1965). Self- and cross-fertilisation in *Spirobis borealis* and *S. pagenstecheri. J. mar. biol. Ass. U.K.*, **45**, 275–285.

Gelfant, S. and Smith, J.C. (1972). Ageing: non-cycling cells, an explanation. *Science, N.Y.*, **178**, 357–361.

Gerking, S.D. (1959). Changes accompanying ageing in fishes. In: *CIBA Foundation Colloquia on Ageing*, **5**, eds. Wolstenholme, G.E.W. and O'Connor, M. Churchill, London, pp. 181–211.

Gershon, H. and Gershon, D. (1973). Inactive enzyme molecules in ageing mice: liver aldolase. *Proc. natn. Acad. Sci.*, **70**, 909–913.

Giese, A. (1973). *Cell Physiology*. Saunders, London.

Giese, A.C. and Pearse, J.S. (1975). *Reproduction of Marine Invertebrates* (vols. 1–3). Academic Press, London.

Glücksman, A. (1951). Cell death in normal vertebrate ontogeny. *Biol. Rev.*, **26**, 59–86.

Goel, N.S. and Yčas, M. (1975). The error catastrophe hypothesis with reference to aging and the evolution of the protein synthesising machinery. *J. theor. Biol.*, **55**, 245–282.

Goldstein, S. (1971). The biology of aging. *New. Engl. J. Med.*, **285**, 1120–1129.

Goldstein, S., Niewiarowski, S. and Sengal, D.P. (1975). Pathological implications of cell ageing *in vitro. Fedn. Proc. Fedn. Am. Socs. exp. Biol.*, **34**, 56–63.

Gonda, O. and Quastel, J.H. (1962). Effects of ouabain on cerebral metabolism and transport mechanisms *in vitro. Biochem. J.*, **84**, 394–406.

Goss, J.A. (1973). *Physiology of Plants and Their Cells*. Pergamon Press, New York, Oxford.

Goss, R.J. (1964). *Adaptive Growth*. Logos Press, London.

Gotto, R.V. (1962). Egg number and ecology in commensal and parasitic copepods. *Ann. Mag. nat. Hist.*, **13**, 97–104.

Green, J. (1972). Vitamin E and the biological antioxidant theory. *Ann. N.Y. Acad. Sci.*, **203**, 29–44.

Hadzi, P. (1910). *Verd. dt. Internat. Kongr. Zool. (Graz)*, **8**. Cited in: Needham, J. (1950). *Biochemistry and Morphogenesis*. Cambridge University Press, Cambridge.

Haemerling, J. (1924). Die ungeschlechtliche Fortzpflanzung und regeneration bei *Aelosoma hemprichii. Zool. Jb.*, **41**, 581–656.

Hahn, H.P. von. (1970). The regulation of protein synthesis in the aging cell. *Expl. Geront.*, **5**, 323–334.

Hall, D.A. (1976). *Ageing of Connective Tissue*. Academic Press, London.

Hamilton, J.B., Terada, H. and Mestler, G. (1955). Studies of growth throughout the life-span in Japanese; Growth and size of nails and their relationship to age, sex, heredity and other factors. *J. Geront.*, **10**, 401–415.

Hamilton, W.D. (1966). The moulding of senescence by natural selection. *J. theor. Biol.*, **12**, 12–45.

Hamilton, W.D. (1972). Altruism and related phenomena mainly in social insects.
    A. Rev. Ecol. Syst., **3**, 193–232.
Haranghy, L. and Balázs, A. (1964–66). Ageing and rejuvenation in planarians.
    Expl. Geront., **1**, 77–91.
Harman, D. (1972). Free radical theory of ageing: dietary implication. *Am. J. clin.
    Nutr.*, **25**, 839–843.
Harms, J.W. (1949). Altem und Somatod der Zellverbandstiere. *Z. Alltersforsch*,
    5, 73–100.
Harper, J.L. (1978), *Population Biology of Plants*, Academic Press, London and New York.
Harper, J.L., Lovell, P.H. and Moore, K.G. (1970). The shapes and sizes of seeds.
    A. Rev. Ecol. Syst., **1**, 327–356.
Harper, J.L. and White, J. (1974). The demography of plants. *A. Rev. Ecol. Syst.*,
    5, 419–463.
Heath, D.J. (1977). Simultaneous hermaphroditism; cost and benefit. *J. theor. Biol.*,
    64, 363–373.
Heinrich, B. and Raven, P.H. (1972). Energetics and pollination ecology. *Science, N.Y.*,
    176, 597–602.
Hemmingsen, A. (1960). Energy metabolism as related to body size and its respiratory
    surface and its evolution. *Rep. Steno. Mem. Hosp. Nord. Insulinlab.*,
    9, 7–110.
Hickman, J.C. (1975). Environmental unpredictability and plastic energy allocation
    in the annual *Polygonum cascadense* (Polygonaceae). *J. Ecol.*, **63**, 689–701.
Hill, A.V. (1950). The dimensions of animals and their muscular dynamics.
    Sci Prog. Oxf., **38**, 209–230.
Hinton, H.E. (1968). Reversible suspension of metabolism and the origins of life.
    Proc. R. Soc., **171B**, 43–57.
Hirshfield, M.F. and Tinkle, D.W. (1975). Natural selection and the evolution of
    reproductive effort. *Proc. natn. Acad. Sci. U.S.A.*, **72**, 2227–2231.
Hoffer, A. and Roy, R.M. (1975). Vitamin C decreases erythrocyte fragility after
    whole body irradiation. *Radiat. Res.*, **61**, 439–443.
Hoffman, G.W. (1974). On the origin of the genetic code and the stability of the
    translation apparatus. *J. molec. Biol.*, **86**, 349–362.
Holland, J.J., Kohn, D. and Doyle, M.V. (1973). Analysis of virus replication in
    ageing human fibroblast cultures. *Nature, Lond.*, **245**, 316–319.
Hollander, C.F. (1970). Functional and cellular aspects of organ ageing. *Expl. Geront.*,
    5, 313–321.
Holliday, R. (1975). Testing the protein error theory of ageing: A reply to Baird,
    Samis, Massie and Zimmerman. *Gerontology*, **21**, 64–68.
Holliday, R. and Tarrant, G.M. (1972). Altered enzymes in ageing human fibroblast
    cultures. *Nature*, (Lond). **238**, 26–30.
Horder, T. (1976). Pattern formation in animal embryos. In: *The Developmental
    Biology of Plants and Animals*, eds. Graham, C.F. and Wareing, P.F.
    Blackwells, Oxford, pp. 169–196.
Houck, J.C. (1973). General Introduction to the Chalone concept. In: *Chalones:
    concepts and current researches.* Natn. Cancer Inst. Monogr., **35**, 1–4.

Huennekens, F.M. (1960). *In vitro* ageing of erythrocytes. In: *Biology of Ageing,*
    ed. Strehler, B.L. Publ. No. 6., Am. Inst. Biol. Sci., Washington D.C.
Hursh, J.B. and Casarett, G.W. (1956). The lethal effect of acute X-radiation on rats
    as a function of age. *Br. J. Radiol.,* **29**, 169–171.
Huxley, J.S. (1926). *Pubbl. Staz. zool. Napoli,* **7**, cited in: Needham, J. (1950).
    Biochemistry and Morphogenesis. Cambridge University Press, Cambridge.
Hyman, L.H. (1919). Physiological studies on planaria I. Oxygen consumption in
    relation to feeding and starvation. *Am. J. Physiol.,* **49**, 377–402.
Hyman, L.H. (1920). Physiological studies on planaria II. A further study of oxygen
    consumption during starvation. *Am. J. Physiol.,* **53**, 399–420.
Jennings, J.B. and Calow, P. (1975). The relationship between high fecundity and the
    evolution of entoparasitism. *Oecologia,* **21**, 109–115.
Johnson, R. and Strehler, B.L. (1972). Loss of genes coding for ribosomal RNA
    in ageing brain cells. *Nature,* (Lond.), **240**, 412–414.
Kauffman, S.A. (1969). Metabolic stability and epigenesis in randomly-constructed
    genetic nets. *J. theor. Biol.,* **22**, 437–467.
Keilin, D. (1953). Stability of biological materials and its bearing on the problem
    of anabiosis. *Scient. Prog.,* **41**, 577–582.
Kempthorne, O. and Pollak, E. (1970). Concepts of fitness in mendelian populations
    Genetics, **64**, 125–145.
Kirkwood, T.B.L. (1977). Evolution of ageing. *Nature,* (Lond.), **270**, 301–304.
Krebs, H.A. and Kornberg, H.L. (1957). *Energy Transformations in Living Matter.*
    Springer, Berlin.
Kleiber, M. (1961). *The Fire of Life.* John Wiley and Sons Inc., New York.
Kleiber, M. (1975). Metabolic turnover rate: a physiological meaning of the metabolic
    rate per unit body weight. *J. theor. Biol.,* **53**, 199–204.
Kohn. H.I. and Kallman, R.F. (1956). Age, growth and the LD_{50} of X-rays. *Science,*
    N.Y., **124**, 1078.
Kohn, R.R. (1971a). *Principles of Mammalian Ageing.* Prentice Hall, Englewood
    Cliffs, New Jersey.
Kohn, R.R. (1971b). Effects of antioxidants on life-span of C57BL mice. *J. Geront.,*
    26, 378–380.
Korenchevsky, V. (1961). *Physiological and Pathological Ageing.* Karger, Basel.
Krohn, P.L. (1962). Heterochronic transplantation in the study of ageing. *Proc. R.*
    Soc., **157B**, 128–147.
Krohn, P.L. (1966). Transplantation and ageing. In: *Topics in the Biology of Ageing,*
    ed. Krohn, P.L. Wiley Interscience, New York, pp. 125–139.
Lange, C.S. (1966). A possible explanation in cellular terms of the physiological
    ageing of the planarians. *Expl. Geront.,* **3**, 219–230.
Lange, C.S. (1967). Studies on the cellular basis of radiation lethality. *Int. J. Radiat.*
    Biol., **13**, 511–530.
Lawlor, L.R. (1976). Moulting, growth and reproductive strategies in the terrestrial
    isopod, *Armadillidium vulgare. Ecology,* **57**, 1179–1194.
Leblond, C.P. and Cheng, H. (1976). Identification of stem cells in the small intestine
    of the mouse. In: *Stem Cells,* eds. Cairnie, A.B., Lala, P.K. and Osmond, D.G.
    Academic Press, London and New York, pp. 7–31.

Lee, J.C.K. (1971). Effects of partial hepatectomy on rats in two transplantable hepatomas. *Am. J. Pathol.*, **65**, 347–356.

León, J.A. (1976). Life histories as adaptive strategies. *J. theor. Biol.*, **60**, 301–336.

Leong, G.F., Grisham, J.W. and Albright, M.L. (1964). Effect of partial hepatectomy on DNA synthesis and mitosis in heterotopic partial autografts of rat liver. *Cancer Res.*, **24**, 1496–1501.

Lesher, S., Fry, R.J.M. and Kohn, H.I. (1961a). Age and the generation time of the mouse duodenal, epithelial cell. *Expl. Cell Res.*, **24**, 334–343.

Lesher, S., Fry, R.J.M. and Kohn, H.J. (1961b). Influence of age on transit time of cells of mouse intestinal epithelium. *Lab. Invest.*, **10**, 291–300.

Lesher, S. and Sacher, G. (1968). Effects of age on cell proliferation in mouse duodenal crypts. *Expl. Geront.*, **3**, 211–217.

Levin, D.A. (1974). The oil content of seeds: an ecological perspective. *Am. Nat.*, **108**, 193–206.

Lewontin, R.C. (1965). Selection for colonising ability. In: *The Genetics of Colonising Species,* eds. Baker, H.G. and Stebbins, G.L. Academic Press, New York, pp. 79–84.

Linn, S., Kairis, M. and Holliday, R. (1976). Decreased fidelity of DNA polymerase activity isolated from ageing human fibroblasts. *Proc. Natn. Acad. Sci. U.S.A.*, **73**, 2818–2822.

Little, J.B. (1976), Relationship between DNA repair capacity and cellular ageing. *Gerontology*, **22**, 28–55.

Lotka, A.J. (1922). Contribution to the energetics of evolution. *Proc. natn. Acad. Sci. U.S.A.*, **8**, 147–151.

MacArthur, R.H. and Wilson, E.O. (1967). *The Theory of Island Biogeography.* Princeton University Press, Princeton, New Jersey.

Macieira-Coelho, A. and Loria, E. (1974). Stimulation of ribosome synthesis during retarded ageing of human fibroblasts by hydrocortisone. *Nature* (Lond.), **251**, 67–69.

Macieira-Coelho, A., Loria, E. and Berumen, L. (1975). Relationship between cell kinetic changes and metabolic events during cell senescence *in vitro*. In: *Cell Impairment, Ageing and Development,* eds. Cristofalo, V.I. and Holěckova, E. Plenum Press, New York.

Maizels, M., Remington, M. and Truscoe, R. (1958). Metabolism and sodium transfer of mouse ascites tumour cells. *J. Physiol.*, **140**, 80–93.

Marsland, D., Landau, J. and Zimmerman, A. (1953). Adenosine triphosphate as an energy source in cell division; pressure-temperature experiments on the cleaving eggs of *Abracia* and *Chaetopterus. Biol. Bull.*, **105**, 366–367.

Martin, G.M., Sprague, C.A., Norwood, T.H., Pendergrass, W.R., Bernstein, P., Hoehn, H. and Arend, W.P. (1975). Do hyperplastoid cell lines differentiate themselves to death? In: *Cell Impairment, Ageing and Development,* eds. Cristofalo, V.I. and Holěckova, E. Plenum Press, New York.

Marx, J.L. (1973). Restriction enzymes: new tools for studying DNA. *Science, N.Y.*, **108**, 482–484.

Mattern, M.R. and Cerutti, P.A. (1975). Age-dependence of excision repair of
γ-ray-damaged thymine by isolated nuclei from diploid human fibroblasts
WI-38. *Nature,* (Lond.), **254**, 450-452.

Maynard Smith, J. (1962). The causes of ageing. *Proc. R. Soc.*, **157B**, 128–147.

Maynard Smith, J. (1964). Kin selection and group selection. *Nature, Lond.*, **201**,
1145–1147.

Maynard Smith, J. (1971a). The origin and maintenance of sex. In: *Group Selection*,
ed. Williams, G.C. Aldine, Chicago, pp. 163–175.

Maynard Smith, J. (1971b). What use is sex? *J. theor. Biol.*, **30**, 319–335.

Maynard Smith, J. (1975). *The Theory of Evolution*. 3rd ed. Penguin Books.

Maynard Smith, J. (1977a). Parental investment: a prospective analysis. *Anim. Behav.*,
25, 1–9.

Maynard Smith, J. (1977b). Why the genome does not congeal. *Nature,* (Lond.),
288, 693–696.

McClendon, J.H. (1975). Efficiency. *J. theor. Biol.*, **49**, 213–218.

McNaughton, S.J. (1975). '*r*' and '*K*' selection in Typha. *Am. Nat.*, **109**, 251–261.

Meats, A. (1971). The relative importance to population increase of fluctuations
in mortality, fecundity and the time variables of the reproductive schedule.
Oecologia, **6**, 223–237.

Mettler, L.E. and Gregg, T.G. (1969). *Population Genetics and Evolution*. Prentice-
Hall, Englewood Cliffs, New Jersey.

Medawar, P.B. (1952), *An unsolved Problem in Biology*. Lewis, London.

Medvedev, Zh. A. (1972). Repetition of molecular-genetic information as a possible
factor in evolutionary changes of life span. *Expl. Geront.*, **7**, 227–238.

Meyerhof, O. (1924). *Chemical Dynamics of Life Phenomena*. Lippincott,
Philadelphia and London.

Mitchison, J.M. (1958). The growth of single cells. II. *Saccharomyces cerevisiae*.
Expl. Cell Res., **15**, 214–221.

Morgulis, S. (1923). *Fasting and Undernutrition*. Dutton, New York.

Morowitz, H.J. (1968). *Energy Flow in Biology*. Academic Press, New York.

Müller, H.J. (1932). Some genetic aspects of sex. *Am. Nat.*, **66**, 118–138.

Müller, H.J. (1949). The Darwinian and modern conceptions of natural selection.
Proc. Am. Phil. Soc., **93**, 459–470.

Murphy, G.I. (1968). Pattern in life-history and the environment. *Am. Nat.*, **102**,
390–404.

Natori, Y. (1975). The effect of ATP on protein degredation in rat liver lysosomes.
In: *Intracellular Protein Turnover*, eds. Schimke, R.T. and Katanuma, N.
Academic Press, London and New York, pp. 237–248.

Needham, A.E. (1962). Is there a second childhood? *Gerontologist*, **2**, 9–13.

Needham, J. (1931). *Chemical Embryology*, vol. 2. Cambridge University Press,
Cambridge.

Newell, R.C. and Pye, V.I. (1971). Variations in the relationship between oxygen
consumption, body size and summated tissue metabolism in the winkle,
Littorina littorea. *J. mar. biol. Ass. U.K.*, **51**, 315–338.

Nichols, J.D., Conley, W., Batt, B. and Tipton, A.R. (1976). Temporally dynamic reproductive strategies and the concept of '*r*' and '*K*' selection. *Am. Nat.*, **110**, 995–1005.

Odum, E.P., Marshall, S.G. and Marples, T.G. (1965). The calorific content of migrating birds. *Ecology*, **46**, 901–904.

Orgel, L.E. (1970). The maintenance and accuracy of protein synthesis and its relevance to ageing, a correction. *Proc. Natn. Acad. Sci. U.S.A.*, **67**, 1476.

Orgel, L.E. (1973). Ageing of clones of mammalian cells. *Nature*, (Lond.), **243**, 441–445.

Packer, L. and Smith, J.R. (1974). Extension of the life-span of cultured, normal human diploid cells by vitamin E. *Proc. Natn. Acad. Sci. U.S.A.*, **71**, 4763–4767.

Parnas, H. and Cohen, D. (1976). The optimal strategy for the metabolism of reserve materials in micro-organisms. *J. theor. Biol.*, **56**, 19–55.

Paul, J. (1965). Carbohydrate and energy metabolism. In: *Cells and Tissues in Culture*, **1**, ed. Willmer, E.N. Academic Press, London and New York, pp. 329–376.

Pedersen, K.J. (1956). On the oxygen consumption of *Planaria vitta* during starvation, the early phases of regeneration and asexual reproduction. *J. exp. Zool.*, **131**, 123–136.

Penning de Vries, F.W.T., Brunsting, A.H.M. and Laar, H.H. Van. (1974). Products, requirements and efficiency of biosynthesis, a quantitative approach. *J. theor. Biol.*, **45**, 339–377.

Phillipson, J. (1966). *Ecological Energetics.* Edward Arnold, London.

Pianka, E.R. (1970). On '*r*' and '*K*' selection. *Am. Nat.*, **104**, 592–597.

Pianka, E.R. and Parker, W.S. (1975). Age-specific reproductive tactics. *Am. Nat.*, **109**, 453–464.

Prescott, D.M. (1955). Relations between cell growth and cell division. I. Reduced weight, cell volume, protein content and nuclear volume of *Amoeba proteus* from division to division. *Expl. Cell Res.*, **9**, 328–337.

Price, G.B. and Makinodan, T. (1973). Ageing: alteration of DNA protein information. *Gerontologia*, **19**, 58–70.

Priede, I.G. (1977). Natural selection for energetic efficiency and the relationship between activity level and mortality. *Nature* (Lond.), **267**, 610–611.

Rabinovich, J.E. (1974). Demographic strategies in animal populations: a regression analysis. In: *Tropical Ecological Systems,* eds. Golley, F.B. and Medina, E. Springer-Verlag, New York, pp. 19–40.

Rendle, A.B. (1930). *The Classification of Flowering Plants,* 2nd ed. Cambridge University Press, Cambridge.

Ricklefs, R.E. (1977). On the evolution of reproductive strategies in birds. *Am. Nat.*, **111**, 453–478.

Rosen, R. (1967). *Optimality Principles in Biology.* Butterworths, London.

Rubinson, H., Kahn, A., Boivin, P., Schapira, F., Gregori, C. and Dreyfus, J.C. (1976). Ageing accuracy of protein synthesis in man: search for inactive enzymatic, cross-reacting material in granulocytes of aged people. *Gerontology*, **22**, 438–448.

Russell-Hunter, W.D. (1961). Life cycles of four freshwater snails in limited popula-
    tions in Loch Lomond, with a discussion on infraspecific variation.
    Proc. Zool. Soc. Lond., **137**, 135–171.

Ryan, J.M., Duda, G. and Cristofalo, V.J. (1974). Error accumulation and ageing
    in diploid cells. *J. Geront.,* **29**, 616–621.

Sager, R. and Ryan, F.J. (1961). *Cell Heredity, An Analysis of the Mechanisms of
    Heredity at the Cellular Level.* Wiley and Sons, New York.

Saunders, J.W. (1966). Death in embryonic systems. *Science, N.Y.,* **154**, 604–612.

Saunders, J.W. (1970). *Patterns and Principles of Animal Development.* Collier-
    MacMillan, London.

Schaechter, M., Williamson, J.P., Hood, J.R. and Koch, A.L. (1962). Growth, cell
    and nuclear division in some bacteria. *J. gen. Microbiol.,* **29**, 421–434.

Schaffer, W.M. (1974). Selection for optimal life histories; the effects of age structure.
    Ecology, **55**, 291–303.

Schimke, R.T. (1975). On the properties and mechanisms of protein turnover. In:
    Intracellular Protein Turnover, eds. Schimke, R.T. and Katanuma, N.
    Academic Press, London and New York.

Schimke, R.T. and Katanuma, W. (eds) (1975). *Intracellular Protein Turnover.*
    Academic Press, London and New York.

Schindler, D.W., Clark, A.S. and Gray, J.R. (1971). Seasonal calorific values as
    determined with a Phillipson Bomb Calorimeter modified for small samples.
    J. Fish. Res. Bd. Can., **28**, 559–564.

Schoenheimer, R. (1946). *The Dynamic Steady State of Body Constituents.*
    Harvard University Press, Cambridge, Mass.

Schroeder, L. (1977). Distribution of caloric densities among larvae feeding on black
    cherry leaves. *Oecologia,* **29**, 219–222.

Schweitzer, P. and Bodenstein, D. (1975). Aging and its relation to shell growth and
    differentiation in *Drosophila* imaginal cells. *Proc. natn. Acad. Sci. U.S.A.,*
    72, 4674–4678.

Sheldrake, A.R. (1974). The ageing, growth and death of cells. *Nature,* **250**,
    381–384.

Simms, H.S. and Berg, B.N. (1957). Longevity and the onset of lesions in the male
    rat. *Gerontologia,* **12**, 244–250.

Simpson, G.G. (1953). *The Major Features of Evolution.* Harcourt, Brace, and
    World Inc., New York.

Sinclair, W.R. and Ross, D.W. (1969). Modes of Growth in mammalian cells.
    Biophysics, **9**, 1056–1070.

Slobodkin, L.B. (1962). Energy in animal ecology. *Adv. Ecol. Res.,* **1**, 69–101.

Slobodkin, L.B. and Richman, S. (1961). Calories/gm in species of animals. *Nature*
    (Lond.), **191**, 299.

Smith, C.G. and Fretwell, S.D. (1974). The optimal balance between size and
    number of offspring. *Am. Nat.,* **108**, 499–506.

Soll. D.R. and Sonneborn, D.R. (1971). Zoospore germination in *Blastocladiella emersonii*: cell differentiation without protein synthesis. *Proc. Natn. Acad. Sci. U.S.A.*, **68**, 459–463.

Soltesz, G., Mestyan, T., Schultz, K. and Rubecz, I. (1973). Glucose disappearance rates and changes in plasma nutrients in normoglycaemic and hypoglycaemic underweight newborns. *Biol. Neonate*, **23**, 139–150.

Sonneborn, T.M. (1930). Genetic studies on *Stenostomum incaudatum* n. sp. I. The nature and origin of differences in individuals formed during vegetative reproduction. *J. exp. Zool.*, **57**, 57–108.

Southwood, T.R.E. (1976). Bionomic strategies and population parameters. In: *Theoretical Ecology*, ed. May, R. Blackwells, Oxford.

Stearns, S.C. (1976). Life-history tactics: a review of ideas. *Q. Rev. Biol.*, **51**, 3–47.

Stearns, S.C. (1977). The evolution of life history traits: a critique of the theory and a review of the data. *A. Rev. Ecol. Syst.*, **8**, 154–171.

Steemann Nielsen, E. and Jørgensen, E.G. (1962). The adaptation to different light intensities in *Chlorella vulgaris* and the time dependence on transfer to a new light intensity. *Physiologia Pl.*, **15**, 505–517.

Steward, F.C. and Krikorian, A.D. (1971). *Plants, Chemicals and Growth.* Academic Press, New York.

Street, P. (1974). *Animal Reproduction.* David and Charles, London.

Strehler, B.L. (1959). Origin and comparison of the effects of time and high-energy radiations on living systems. *Q. Rev. Biol.*, **34**, 117–142.

Strehler, B.L. (1962). *Time Cells and Ageing.* Academic Press, London and New York.

Strehler, B.L. (1975). Implications of ageing research for society. *Fedn. Proc. Fedn. Am. Socs. Exp. Biol.*, **34**, 5–8.

Strehler, B.L. and Mildvan, A.S. (1960). General theory of mortality and ageing. *Science, N.Y.*, **132**, 14–21.

Strong, L.C. (1968). *Biological Aspects of Cancer and Ageing.* Pergamon Press, Oxford.

Strobeck, C., Maynard Smith, J. and Charlesworth, B. (1976). The effects of hitchhiking on a gene for recombination. *Genetics*, **82**, 547–558.

Tanner, J.M. (1963). Regulation of growth in size in mammals. *Nature* (Lond.), **199**, 845–850.

Tappel, A.L. (1968). Will antioxidant nutrients slow ageing processes. *Geriatrics*, **23**, 97–105.

Tappel, A.L., Fletcher, B. and Deamer, D. (1973). Effect of antioxidants and nutrients on lipid peroxidation, fluorescent products and ageing parameters in the mouse. *J. Geront.*, **28**, 415–434.

Tas, S. (1977). Note on the Orgel hypothesis and on the mechanism of aging. *Gerontology.* **23**, 306–308.

Teal, J.M. (1957). Community metabolism in a temperate cold spring. *Ecol. Monogr.*, **23**, 41–78.

Thompson, R.C. and Ballou, J.E. (1956). Studies of metabolic turnover with tritium as a tracer. V. The predominantly non-dynamic state of body constituents in the rat. *J. biol. Chem.*, **223**, 795–804.

Thrasher, J.D. (1967). Age and cell cycle of the mouse colonic epithelium. *Anat. Rec.*, **157**, 621–626.

Tilley, S.G. (1968). Size-fecundity relationships and their evolutionary implications in five desmognathine salamanders. *Evolution*, **22**, 806–816.

Timaras, P.S. (1975). Ageing of homoestatic control systems. *Fedn. Proc. Fedn. Am. Socs. Exp. Biol.*, **34**, 81–82.

Tinkle, D.W. (1969). The concept of reproductive effort and its relation to the evolution of life-histories of lizards. *Am. Nat.*, **103**, 501–516.

Tomkins, G.A., Stainbridge, E.J. and Hayflick, L. (1974). Viral probes of ageing in the human diploid cell strain WI-38. *Soc. exp. Biol. Med.*, **146**, 385–390.

Tomlinson, J. (1966). The advantage of hermaphroditism and parthenogenesis. *J. theor. Biol.*, **11**, 54–58.

Treisman, M. (1976). The evolution of sexual reproduction, a model which assumes individual selection. *J. theor. Biol.*, **60**, 421–431.

Treisman, M. and Dawkins, R. (1976). The 'cost of meiosis', is there any? *J. theor. Biol.*, **63**, 479–484.

Twitty, V.C. (1940). Size controlling factors. *Growth* suppl. Second Symp. on Development and Growth.

Verzar, F. (1963). The ageing of collagen. *Sci. Am.*, **208**, 104–114.

Wassink, E.C., Kok, B. and van Oorschot, L.P. (1964). The efficiency of light energy conversion in *Chlorella* cultures as compared with higher plants. In: *Algal Culture*, ed. Barrow, J.S. Carnegie Inst. Publ. **600**, Washington, D.C. pp. 55–62.

Waterlow, J.C. (1967). Lysine turnover in man measured by intravenous infusion of L-[u–^{14}C] lysine. *Clin. Sci.*, **33**, 507–515.

Waterlow, J.C. and Stephen, J.M.L. (1967). The measurement of total lysine turnover in the rat by intravenous infusions of L-[u–^{14}C] lysine. *Clin. Sci.*, **33**, 489–506.

Watson, J.D. (1970). *Molecular Biology of the Gene*. W.A. Benjamin Inc., California and London.

Weismann, A. (1891). The duration of life. In: *Essays upon heredity*, Oxford University Press, Oxford.

Weiss, P. and Hiscoe, H.B. (1948). Experiments on the mechanism of nerve growth. *J. exp. Zool.*, **107**, 315–395.

Weiss, P. and Kavanau, J.L. (1957). A model of growth and control in mathematical terms. *J. gen. Physiol.*, **41**, 1–47.

Welch, H.E. (1968). Relationship between assimilation efficiencies and growth efficiencies for aquatic consumers. *Ecology*, **49**, 755–759.

Wendt, H. (1965). *The Sex Life of Animals*. Arthur Baker Ltd., London.

Wheeler, K.T. and Lett, J.T. (1974). On the possibility that DNA repair is related to age in non-dividing cells. *Proc. Natn. Acad. Sci. U.S.A.*, **71**, 1862–1865.

Wiegert, R.G. and Coleman, D.C. (1970). Ecological significance of low oxygen consumption and high fat accumulation by *Nasutermes costalis* (Isoptera: Fermitidae). *Bioscience*, **20**, 663–665.

Williams, G.C. (1957). Pleiotoropy, natural selection and the evolution of senescence. *Evolution,* **11**, 398–411.

Williams, G.C. (1966a). *Adaptation and Natural Selection.* Princeton University Press, Princeton, New Jersey.

Williams, G.C. (1966b). Natural selection, the cost of reproduction and a refinement of Lack's principle. *Am. Nat.,* **100**, 687–692.

Williams, G.C. (1975). *Sex and Evolution.* Princeton University Press, Princeton.

Williams, G.C. and Mitton, J.B. (1973). Why reproduce sexually? *J. theor. Biol.,* **39**, 545–554.

Winterer, J.C., Steffee, W.P., Davy, W., Perera, A., Scrimshaw, N.S. and Young, V.R. (1976). Whole body protein turnover in ageing man. *Expl. Geront.,* **11**, 79–87.

Wissing, J.E., Darnell, R.M., Ibrahim, M.A. and Berner, L. (1973). Calorific values of marine animals from the Gulf of Mexico. *Contr. in mar. Sci.,* **17**, 1–7.

Wolpert, L. (1969). Positional information and spatial pattern of cellular differentiation. *J. theor. Biol.,* **25**, 1–47.

Wolpert, L. (1970). Developing cells know their place. *New Scientist,* **46**, 322–325.

Woolhouse, H.W. (1972). *Ageing Processes in Higher Plants.* Oxford Biology Readers, **30**, Oxford University Press, Oxford.

Wulf, J. and Cutler, R. (1974). Altered protein hypothesis of mammalian ageing processes. *Expl. Geront.,* **10**, 101–117.

Yagil, G. (1976). Are altered glucose-6-phosphate dehydrogenase molecules present in aged liver cells? *Expl. Geront.,* **11**, 73–78.

Young, J.Z. (1971). *An Introduction to the Study of Man.* Oxford University Press, Oxford.

Young, V.R. (1976). Protein metabolism and needs of elderly people. In: *Nutrition and Ageing,* eds. Rochstein, M. and Sussman, M.L. Academic Press, New York, pp. 67–102.

Yousef, M.K. and Johnson, H.D. (1970). Effects of heat, diet and hormones on total body protein turnover in young and old rats. *Proc. Soc. exp. Biol. Med.,* **135**, 765–766.

Yushok, W. (1974). ATP-dependent turnover of tumor cell proteins modified by amino acid autolysis. *Fedn. Proc., Fedn. Am. Socs. exp. Biol.,* **33**, 1545.

Zeman, W. (1971). The neuronal ceroid-lipofuscinosis-Batton-Vogt syndrome: A model for ageing. In: *Advances in Gerontological Research,* ed. Strehler, B.L. Academic Press, New York, pp. 147–170.

Zeuthen, E. (1947). Body size and metabolic rate in the animal kingdom. *Cr. Trav. Lab. Carlsberg, Ser. Chem.,* **26**, 15–161.

Zeuthen, E. (1950a). Cartesian diver respirometer. *Biol. Bull.,* **98**, 139–143.

Zeuthen, E. (1950b). *Respiration during cell division in the egg of the sea urchin. Psammechinus mitiaris. Biol. Bull.,* **98**, 144–151.

Zeuthen, E. (1950c). Respiration and cell division in the egg of *Urechis caupo. Biol. Bull.,* **98**, 152–156.

Zotin, A.I. (1972). Thermodynamic aspects of developmental biology. In: *Monographs in Developmental Biology,* **5**, ed. Wolsky, A. Karger, S., Basel, München, Paris, London, New York, Sydney.

Index of organisms

Only organisms mentioned in the text appear in the list; they are grouped into general taxonomic categories and then arranged alphabetically.

PROKARYOTES (without nuclear membrane), 13–15

    VIRUS, 104

    BACTERIA, 14, 16, 23, 27, 28, 43, 51, 75

        Escherichia coli, 41, 43, 177

        Pneumococcus, 14

    BLUE-GREEN ALGAE, 57

EUKARYOTES (with nuclear membrane), 13–15

PROTISTA (unicellular plants and animals), 16

    ALGAE (except blue-green; autotroph), 29, 68

        Chlamydomonas reinhardi, 77

        Chlorella vulgaris, 35, 36

    PROTOZOA (heterotroph), 28

        Amoeba proteus, 41

        Paramecium, 69

        Volvox (colonial), 17

ANIMAL KINGDOM

    PLACOZOA (multicellular, very loosely organized), 17

    PARAZOA (multicellular, loosely organized)

    PORIFERA (sponges), 17

    METAZOA (multicellular, much organization and specialization) 17, 84, 111, 124

    COELENTERATA (= *CNIDARIA*), 17, 84, 111, 124

        Hydra, 17, 44, 67

        Coral, 78

    PLATYHELMINTHES

        Turbellaria (mainly free-living), 32

        Rhabdocoela

        Stenostomum incaudatum, 133

152

Tricladida (flatworm or planarian), 17, 37, 62, 63, 103, 124–133
    Crenobia alpina, 125
    Dendrocoelum lacteum, 125, 128, 131
    Dugesia polychroa, 125
    Dugesia lugubris, 37, 123
    Planaria torva, 125
    Polycelis felina, 125
    Polycelis nigra, 125
    Polycelis tenuis, 125, 126, 128
Temnocephalida (ectosymbiotic), 62
Trematoda (entosymbiotic flukes), 62, 66, 67
Cestoda (entosymbiotic tapeworms), 62
ASCHELMINTHES
Nematoda (round worms), 38, 110, 111, 113, 119
Rotifera (wheel animalicules), 77, 110, 113
NEMERTINA (ribbon worms), 121
ANNELIDA (true worms), 17, 43, 111
    Polychaeta
      Serpulidae, 133
    Oligochaeta
      Aeolosoma, 133
MOLLUSCA
    Gastropoda (snails), 104
    Pulmonates
      Ancylus fluviatilis, 92, 105, 106, 109
      Lymnaea palustris, 92
      Lymnaea pereger, 92, 105, 106
      Lymnaea stagnalis, 92
      Physa fontinalis, 92
      Physa gyrina, 92
      Planorbis contortus, 34, 92, 105, 106, 108, 109
    Prosobranchs
      Bithynia tentaculata, 92
    Lamellibranchia (bivales) Oyster, 72–78
    Cephalopoda (squid, octopus), 84
TARDIGRADA (bear animalicules), 38
ARTHROPODA
    Trilobita (extinct), 57
    Diplopoda (millipedes), 99
      Cylindroiulus latestriatus, 99
      Cylindroiulus punctatus, 99
      Iulus scandinavius, 99
      Ophyiulus pilosus, 99

Crustacea
Isopods
  Armadillidum vulgare, 93
Copepods, 104
Insecta, 17, 61, 72, 110, 113
  Aphids, 77
  Bees, 75
  Biston betularia (peppered moths), 10
  Drosophila, 15, 71, 76, 130
  Drosophila melanogaster, 77
  Fireflies, 72
  Termites, 61
Merostomata
  Eurypterids (extinct), 57
Arachnida (spiders), 33
ECHINODERMATA
  Sea urchin, 43, 72
CHORDATA
  Urochordata
    Ascidians, 124
    Appendicularia (Larvacea), 38
  Protochordata, 79
  Vertebrata, 82, 110
  Teleostei, (bony fish), 30, 31, 111
    Loach, 23
    Stickleback, 70
  Amphibia, 17
    Frogs, 71
    Salamanders, 59
    Xenopus laevis, 40
  Reptilia, 111
    Diplodocus (extinct), 57
    Ichthyosaurus (extinct), 57
    Lizards, 93–95
  Aves (birds), 52, 61, 71, 72, 75, 79, 95, 111
  Mammalia, 72, 79
    Homo sapiens, 77
    Mus musculus, 77, 110
    Rodents, 120
    Whales, 56, 57, 61

PLANT KINGDOM
  EUMYCOPHYTA (fungi), 28, 68, 104
    OOMYCETES (water moulds)
        Blastocladiella emersonii, 43
    ASCOMYCETES (sac fungi)
        Neurospora crassa, 77
        Yeast, 28
  TRACHEOPHYTA (vascular plants)
    LYCOPSIDA (club mosses), 57
    PTEROPSIDA (ferns), 73
    SPERMOPSIDA (seed plants)
      Coniferae, 57
      Angiospermae
        Acer, 97
        Barley, 92
        Betula, 97
        Centopogen talamencis, 74
        Dandelions, 98
        Erythina umbrosa, 75
        Erythina subumbrans, 73
        Elm, 78
        Grasses, 72
        Maize, 92
        Meadow fescue, 74
        Quercus, 97
        Sedges, 94
        Sequoia, 56
        Solidago canadensis, 96, 97
        Solidago nemoralis, 96, 97
        Solidago rugosa, 96, 97
        Solidago speciosa, 96, 97
        Zea mays, 77

Subject index

Abrasion, 113
Accident, 93, 107, 113
Acer, 97
Acid phosphatase, 51, 53
Active transport, 22, 28
Adaptation, 56, 114
Adaptability, 33, 35, 100
Adaptive value, 10
Adenine, 11
Adiposity, 60–62
ADP, 22, 23
Absorption efficiency, 31–33
Aeolosoma, 133
Aerobic respiration, 21, 22
Age, 45, 60, 95, 136
   'pigment', 118
   specific mortality, 91
Ageing, 1, 107–136
Albino, 77
Albumen gland, 85
Algae, 29, 61, 68
Allele, 8–10, 15, 72, 79
Allometric, 129
Allosteric, 13
Altruism, 10
Amino acid, 12, 13, 30, 49
   analogue, 50
Amoeba proteus, 41
AMP, 23
Amphibia, 17, 57, 71, 111
Anabolism, 126
Anaerobic respiration, 21, 22
Ancylus fluviatilis, 92, 105, 106

Animal Kingdom, 21, 69, 84
Annelida, 17, 43, 111
Annual, 112
Analogue, 50, 117
Antioxidant, 119
Aphid-rotifer model, 77
Appendicularia, 38
Appendix, 63
Appetite, 35
Armadillidum vulgare, 93
Arthropoda, 111
Ascidians, 124
Assortment, genetic, 80
ATP, 21–28, 49, 51
Attraction, 71
Autotroph, 21
Autoxidation, 119
Auxin, 55
Axon, 50

β-galactosidase, 15
Bacteria, 14, 16, 23, 27, 28, 43, 51, 75
Barley, 92
Basal metabolism, 30
Bees, 75
Bertalanffy growth model, 126
Betula, 97
Binary fission, 67
   nets, 135
Biomass, 19, 25, 26, 29, 61
Biosphere, 20
Biotype, 98
Birds, 61, 71, 72, 79, 95, 111
Birth, 95

Biston betularia, 10
Bitch, 73
Bithynia tentaculata, 92
Blastocadiella emersonii, 43
Blastomeres, 42, 43
Blood, 42, 51, 52
Blue-green algae, 57
Bones, 42, 48, 52
Bond, 26
Brain, 48
Breeding season, 88, 94, 101
Budding, 67
Buoyancy, 61

Calorific value, 61
Carbohydrate, 21
Carbon, 3, 5
    dioxide, 21
Cardiac index, 110
Carnivore, 31, 33
Carrying capacity, 89
Cartilage, 52
Catabolism, 126
Catalase, 49
Catastrophe, 116, 119, 120
Cecaria, 67
Cell, 13–16, 19, 26–28, 37–39,
        42–46, 50–53
    cycle, 39, 44, 45
    death, 50–53
    division, 39, 43–45, 47, 67, 121;
        see also Meiosis and Mitosis
    growth, 40–41, 55
    population, 50
    standing crop, 39
    wall, 14
Centopogen talamencis, 74
Central dogma, 7, 14, 43
Centromere, 67
Cephalopoda, 84
Cereals, 73
Cestoda, 62
Chastity, 93
Chemical work, 22
Chick, 52
Chlamydomonas reinhardii, 77
Chlorella vulgaris, 35, 36

Chlorophyll, 2, 36
Chloroplast, 14
Chromatids, 67
Chromatin, 12, 119
Chromosome, 14, 66, 67, 76
    aberration, 115, 116
    puff, 14, 15, 42
Clawed toad, 40
Cleavage, 39
Clone, 78
Closed system, 2
Clouds, 3
Co-adaptation, 72
Coefficient of selection, 9, 30, 87
Coelenterata, 17, 84, 111, 124
Collagen, 48
Colonization, 77
Combat, 58
Communication, 6, 11–12, 18
Comparative physiology, 24
Competition, 58, 78
Computer model, 63
Conformer, 101
Conifer, 57
Conjugation, 69
Copepod, 104
Cope's law, 16
Copulation, 70–73, 93
Copying, 66
Cordaites, 57
Corolla, 75
Coronary, 110
Corrosion, 113
Cost, 84, 92
    of meiosis, 75–83
Courtship, 93
Cow, 73
Crenobia alpina, 125
Cross-linkage, 115, 118, 119
Cross-over, genetic, 7, 68, 76, 80
Crystals, 3, 4
Cyanide, 123, 128
Cylindroiulus latestriatus, 99
Cylindroiulus punctatus, 99

Damage, 7, 12, 13, 49, 113–115,
        119, 120

Dandelions, 98
Death, 107—110, 113
Defaecation, 3, 24; *see also* Egestion
Degeneration, 17, 136
Degrowth, 124—133
Demography, 96
Denaturation, 49, 50
Dendrocoelum lacteum, 125, 128, 131
Density-dependent, 96, 98
Density-independent, 96, 98
Dermis, 42
Desertion, 70
Design, 5, 17, 24, 117
Development, 45
Developmental biology, 4
Diakenesis, 68
Differentiation, 16, 41—44, 52, 121,
    122
Dimorphism, sexual, 69
Dioecious, 71
Diploid, 7, 66—68, 83
Diplodocus, 57
Diptera, 42
Disease, 65, 93, 107, 121, 136
    infectious, 110
Dispersal, 104
DNA, 11—15, 38, 39, 42, 115, 116,
    119
Domestication, 108
Dominance, genetic, 9, 20, 31, 69
Donor, 112
Dose-response, 130, 131
Double helix, 11
Drosophila, 15, 71, 76, 130
    melangogaster, 77
Drug, 41
Ducks, 70
Dugesia lugubris, 37, 125
    polychroa, 25
Dynamic steady-state, 19, 47—55

Ecology, 1, 56, 88, 93, 99
Ectoderm, 42
Ectosymbiotes, 62
EDTA, 37
Efficiency, 21, 24—30, 45, 101
    photosynthetic, 29, 35
Egestion, 3, 24; *see also* Defaecation

Egg, 69, 71, 86, 104—106, 134
Elm-oyster model, 78
Embryo, 28, 29, 39, 44, 45, 53, 134
Embryogenesis, 45, 52
Endoderm, 42
Endonuclease, 46
Endoplasmic reticulum, 14
Energy, 2, 5, 6, 19, 22, 24, 25, 37,
    43—45, 56—58, 100—104,
    121
    budget, 30, 63
    charge, 23, 51
Entoparasitic, 92, 100
Entropy, 2, 47, 135
Enzyme, 29, 49, 51, 116, 117
    induction, 117, 118
Epidermis, 42
Equilibrium, 9, 10
Errors, 5, 49, 50, 115, 116
Error catastrophe, *see* Catastrophe
Erythina, 75
    subumbrans, 75
    umbrosa, 75
Escherichia coli, 41, 43, 177
Ester bond, 26
Ethoxyquinone, 119
Eugenes fulgens spectabilis, 75
Eukaryote, 11, 14, 28, 42, 43
Eurypterid, 57
Evolution, 5, 7, 114
Evolutionary biology, 1
Excretion, 30, 33, 42
Exonuclease, 46

Family, 70
Fatty acid, 30
Fecundity, 7, 94
Feeding strategy, 106
Female, 69—71
Fertility, 10
Fertilization, 71—75
Fern, 73
Fibroblast, 51—53, 117
Firefly, 72
Fish, 31, 57, 11
Fitness, 7, 10, 25, 30, 56, 75, 88, 91,
    100, 101, 105, 121, 136

Fixation, genetic, 10, 82
Flagellum, 14, 69
Flatworm, 32
Flukes, 62
Food, absorption of, 24
Free radicals, 118
Frogs, 71
Fruit, 38
Fungi, 68, 104

G phase, 39–44, 121
Gametes, 12, 30, 56, 59, 60, 66–69,
    72–75, 83, 84, 100–104
Gametogenesis, 134
Gametophyte, 69
Gastropoda, 85, 104
Gene, 7–10, 43, 55, 57, 76, 79, 88,
    114, 115
  pool, 61, 83, 88
Generation, 4, 6, 17, 31, 88
Genetic code, 11, 12
  dominance, 10
Genome, 43, 44, 49, 52, 53, 135, 136
Genotype, 5–8, 11, 76, 77, 80–82,
    135, 136
Germ cells, 17, 42, 122
Gibberellin, 51
Glucokinase, 49
Glycolysis, 22
Glycosidic bond, 26
Gompertz equation, 108–110
Gompertz plot, 109, 111
Gonad, 84–86
Gonadectomy, 93
Gonochorism, 83–86
Gradient, 43
Graft, 54
Grain, 92
Grasses, 74
Gross efficiency, 30, 33
Growth, 1, 17–19, 46, 56, 93, 110,
    120–127
Guanine, 11
Gut, 39, 42

Habitat, 70–80, 87
Haeckel, 42

Haemoglobin, 42
Haemophilic, 77
Hair, 120
Half-life, 27, 47, 48
Haploid, 66–69, 83
Hardy-Weinberg theorem, 8
Helium, 3
Hepatectomy, 41, 53
Herbivore, 31, 33
Hermaphrodite, 69, 83, 86
Hermaphrodite duct, 85
Herpes virus, 117
Heterotrophs, 21–29
Heterozygote, 69, 79
  advantage, 10
  disadvantage, 10
Histogenetic cell deaths, 52
'Hitch-hiking', genetic, 83
Holometabolous, 52
Homeostasis, 29, 25, 33
Homeotherm, 28, 29, 61, 104
Homologous, 67
Homo sapiens, 77
Homozygote, 9, 79
Hormone, 16–17, 118
Host, 104, 112
Hummingbird, 75
Hydra, 17, 44, 67
Hydrogen, 3
Hydrogen bond, 11
Hygiene, 108
Hymenoptera, 10
Hyperplasia, 38, 46
Hypertrophy, 38, 46

Ichthyosaur, 57
Immigration, 105
Immortality, 107, 113, 122
Information, 6, 7, 11, 25, 68
Ingestion, 30–35
Inhibition, 33, 55
Innate capacity for increase, 87, 88
Inositol, 77
Insecta, 17, 61, 72, 110, 113
Insulation, 61
Insurance, 62, 91, 92
Invertebrata, 86, 110

Internal combustion engine, 24

Intestine, 48

Interphase, 40

Iron, 5

Irradiation, γ, 129—133
    X, 116

Isometric, 129

Isopoda, 93

Isotope, 47, 51

Isozyme, 98

Iteroparous, 90—92

Iulus scandinavius, 99

'*K*' selection, 89, 95, 96, 104

Keratin, 50

Kidney, 48, 53, 55

Kinin, 55

Lactate dehydrogenase, 49

Lactose, 15, 43

Lamarck, 7, 135

Larvae, 104

Leaves, 97, 122

Ligase, 49

Light, 29, 35

Light intensity, 35

Lipid, 14

Lipofuscin, 118

Liver, 38, 39, 44, 48, 49, 53

Lizards, 93, 95

Loach, 23

Loci, 76, 79, 83

Locomotion, 29, 33, 58

Lung, 48

Lycopod, 57

Lymnaea palustris, 92
    pereger, 92, 105, 106
    stagnalis, 92

Lysosomes, 49, 51, 53

Machines, biological, 24

Macrophyte, 29, 35, 61, 68, 73

Maintenance, 33

Maize, 92

Male, 69—71

Mammal, 72, 79, 111, 116, 130

Material, 2, 5, 19, 37, 54, 65

Mating, 84

Mating call, 72

Maturity, 61, 65, 71, 95

Meadow fescue, 74

Medusa, 124

Medicine, 108

Meiosis, 66—68
    cost of, 76—83

Melanism, 10

Membrane, 13, 28

Mendel, 8

Meristem, 122

Mesenchyme, 42

Mesoderm, 42

Mesonephros, 52

Message, 6, 116

Metal fatigue, 111

Metamorphosis, 53

Metazoa, 16, 41, 46, 65

Mice, 23, 110, 77

Migration, 8, 61

Millipede, 99

Miracidium, 67

Mistake, 120

Mitochondria, 15, 28, 51

Mitosis, 28, 39, 44—46, 66—69,
    121, 132

Molecular biology, 14

Molecule, 50, 51, 115—123

Mollusca, 43, 84, 111

Monomer, 27

Morphogenesis, 55

Morphogenetic cell death, 52

Mortality, 90—96, 104—110

Mosaic development, 43

Mother cell, 134

Mould, 28

mRNA, 13 —15, 43

Mucus, 31—33

Muscle, 39, 42, 48, 50, 121

Mutation, 7, 8, 12, 76, 80—82, 111,
    116

Nail, 120

Natural selection, *see* Selection

Necrosis, 52, 53

Nectar, 74

Nematoda, 38, 110–113, 119
Nermertina, 124
Neoblast, 132–134
Nerves, 17, 27, 39, 42, 45, 50
Net efficiency, 20, 31, 32
Neurospora, 77
Noise, *see* thermal noise
Notochord, 42
Nucleic acid, 11, 28, 66
Nucleotide, 11, 12
Nucleus, 3, 14–15, 51, 68
Nutrient, 46, 56

Obesity, 60, 62
Oligochaeta, 133
Onsager's reciprocal relation, 23
Ontogeny, 1, 42
Oothecal gland, 85
Open system, 2, 20–23
Ophyiulus pilosus, 99
Optimality, 136
Organelle, 28, 118
Organism, 16–18
Organismic biology, 136
Ornithine decarboxylase, 49
Ostrich egg, 14
Ouabain, 27
Ovary, 86
Ovulation, induced, 73
  spontaneous, 73
Ovotestis, 85
Oxygen uptake, 128
Oysters, 72

Pachytene, 68
Paramecium, 69
Parasite, 62, 67, 104
Parental care, 93, 70, 71
Parthenogenesis, 75–80
Parturition, 93
Pathology, 110, 113
Pattern, 43–45
Pelt, 48
Penis, 85
Peppered moth, 10
Peptide bond, 26
Perennial, 92

Persistence, 3–6
Phagocyte, 52, 53
Phenotype, 5–8, 11–13, 77, 135, 136
Pheromones, 72
Phosphate bond, 21, 26
Photosynthesis, 20–29
Phylogenetic cell death, 52
Phylogeny, 1
Physa fontinalis, 92
Physa gyrina, 92
Piebald, 77
Pistil, 73, 74
Pith, 122
Pituitary, 17
Placozoa, 17
Planaria torva, 125, 126
Plankton, 104
Planorbis contortus, 34, 92, 105, 106,
  108, 109
Plans, 4–6
Plasma, 49
Platyhelminthes, 62, 84, 111
Pneumococcus, 14
Poikilotherm, 28, 29, 31, 33, 59
Polio virus, 117
Pollen, 73, 74
Pollination, 74
Polycelis felina, 125
  nigra, 125
  tenuis, 125–128
Polygene, 79
Polymer, 25, 27
Polymerase, 49, 116, 120
Polymeric bond, 26
Polyploidy, 38
Population, 7, 10, 69, 85, 88, 89, 94,
  105
Population density, 89
Population genetics, 8
Population growth, 89
Porifera, 17
Potential energy, 21
Predators, 5, 58, 65, 72, 93, 104, 107,
  121, 136
Pregnancy, 72, 93
Prigogine's theorem, 23
Primary structure, 13

Progeria, 116
Programme, 6, 7, 17, 122, 135
Prokaryote, 14
Proliferation, cell, 39, 54, 55
Pronephros, 52
Prostate gland, 85
Protandry, 86
Protease, 49
Protein, 11–14, 27, 48–50, 119, 121
Protista, 16
Protochordata, 39
Protozoa, 28
Pseudopregnancy, 73
Punctuation, 13
Pupation, 61

Quercus, 97

'r' selection, 95, 96, 104
Rat, 38, 48, 49, 110
Ration, 102, 103
Receptor site, 118
Recessive gene, 69
Reciprocity, 135
Recklessness, reproductive, 102
Recombination, genetic, 82
Recombinational load, 82
Red blood cells, 51, 52
Redia, 67
Redundancy, 13
Regeneration, 17
Regrowth, 126, 131–133
Rejuvenation, 107, 124, 127–129,
        132–134
Rejuvenation, model of, 132
Renal tubules, 44
Repair, 12, 49, 119–123, 135, 136
Replication, 3–7, 11–15
Repression, 44
Reproduction, 1, 17, 18, 66–106,
        122, 123
    asexual, 66, 76–83, 124, 133
Reproductive effort, 89, 93–95, 99,
        100, 104
    restraint, 102
    strategies, 96
    value, 93–96, 101
        residual, 94, 95, 114

Reptilia, 111
Resources, 89
Respiration, 20–22, 28, 128, 129
Respiratory rate, 32, 34
Rhabdocoela, 133
Rheostat, 122
Rheumatoid arthritis, 110
Ribosome, 14
RNA, 11, 13, 42
Rodents, 120
    desert, 102
Roots, 39
Rotifers, 38, 110, 113

S phase, 39, 40, 44
Salamander, 59
Salivary gland, 44
Satiety, 62
Seasonality, 60
Searching, 34
Sea urchin, 43, 72
Secondary structure, 13
Secretion, 33
Sedges, 74
Seed, 38, 104, 106
Selection, 8, 9, 16, 24, 50, 56,
        63–65, 69, 89, 96, 112, 121
    group, 10
    kin, 10, 114
Selective value, 8, 9, 87
Self-fertilization, 85, 86
Semelparous, 90–92
Semen, 72
Senescence, 17, 18
Sequoia, 56
Serpulids, 133
Sex, 7, 58, 66–68, 75–86, 100, 124,
        133
    -linked genes, 8
    ratio, 76
Sexual dimorphism, 69
    selection, 69
Shade plant, 76
Shoots, 39
Sigmoid growth, 127
Signal, 43, 44, 71
Simulation, 96
Sink, 20

Size, 16, 17, 19, 23, 47, 57—60
Skin, 39, 50—53
Snails, 32, 60, 103, 108
Sockeye salmon, 35
Solidago canadensis, 96, 97
    nemoralis, 96, 97
    rugosa, 96, 97
    speciosa, 96, 97
Soma, 122
Somatic mutation, 111, 116
Somite, 42
Source, 20
Specialization, 16
Speed, 85
Sperm, 69, 71, 75, 134
Spermatheca, 85
Spider, 33
Spore, 67
Sporocyst, 67
Stability, 3—5
Starvation, 34, 35, 124—127
Steady-state, 48, 89, 120
    size, 19, 47
Steak, 50
Stem, 97
    cell, 42, 122, 132—134
Stenostomum incaudatum, 133
Stickleback, 70
Stigma, 74
Stimulation, 53, 55
Stochastic, 119, 122
Stomach, 48
Storage, 50, 60, 62, 63
Strategy, 4, 105, 106
Strawberry-coral model, 78
Strength, 58
Streptomycin, 77
Suicide, 113
Sun, 20
Sunbird, 75
Sun plant, 36
Survival, 7, 10, 19, 48, 93—99, 121, 130
Survivorship curve, 108, 109
Suspended animation, 118
Switch, 14, 43, 122, 136
Symbiosis, 15

Synchronization, 112

Tardigrades, 38
TCA cycle, 22
Teleosts, 30
Telophase, 68
Template, 54, 119
Temnocephalids, 62
Temperature, 80
Termite, 61
Thermal noise, 12, 13, 43
Thermodynamics, laws of, 1, 2, 21, 135
Thymine, 11
Thyroid, 52
Tissue, 47, 48
Tissue culture, 119
Tracers, 51
Trade-off, 91
Transcription, 11, 12, 43, 49, 119
Transfer function, 100
Translation, 43, 49
Transplantation, 112
Trees, 92, 112
Treisman model, 79
Trematodes, 66, 67
Triclads, 17, 62—63, 67, 103, 124, 133
Trilobites, 57
Triplet code, 13
Tritiated thymidine, 51
    uridine, 42
Tritium, 48
tRNA, 13
Turbellaria, 17
Turnover, 27, 47, 48, 50—55, 112, 113, 120—122
Twins, 113

Uracil, 13

Van der Waal's force, 11
Vector, 74
Vegetative reproduction, 68
Verhulst-Pearl equation, 89
Vertebrata, 82, 110
Viability, 86
Virus, 104
Vital dye, 52

Vitamin E, 119
Viviparity, 86
Volvox, 17

Weiss-Kavanau model, 54
Welfare state, 115
Whales, 56–57, 61
Wing bud, 52–53
Work, 4, 22
Wounding, 55, 132
WZ system, 71

Xenopus laevis, 40
X–irradiation, 116
XY system, 71

Yeast, 28
Yolk sac, 52

Zea mays, 77
Zygote, 8, 41, 43, 75, 83, 86, 134